YOU MATTER MORE THAN YOU THINK

YOU MATTER MORE THAN YOU THINK

Quantum Social Change for a Thriving World

KAREN O'BRIEN

Foreword by Christina Bethell
Art by Tone Bjordam

C

cCHANGE Press
Oslo, Norway

2021

cCHANGE Press
Oslo, Norway
https://cchange.no/cchangepress
www.youmattermorethanyouthink.com

First published in 2021

ISBN 978-82-691819-3-7 (paperback)
ISBN 978-82-691819-5-1 (hardcover)
ISBN 978-82-691819-4-4 (e-Pub)

All artwork and Figure 1 are copyrighted by Tone Bjordam (www.tonebjordam.com)
The poem "Connection" is copyrighted by Shohini Ghose.

Interior design and formatting: Qamber Designs (qamberdesigns.com)
Cover artwork: Tone Bjordam
Cover design: Heidi Bragerhaug (bravoo.no)

PRINTED IN
NORWAY
NO - 1650

MILJØMERKET
241   450
Trykksak

This book is for you.

We actually live in a quantum world, and once we fully grasp that, nothing will ever be the same again.

—Danah Zohar

CONTENTS

FOREWORD

THE IMPORTANCE OF MATTERING

"I see you, you are important to me and I value you."
—*Sawubona*, an African greeting

Like in quantum physics, when we peer deeply into the science of human flourishing we find at its roots our fundamental interconnectedness. As children or adults, we experience our innate interconnectedness through a felt sense of belonging and mattering, which requires that we feel seen, valued, and shown we are important to others and the web of life in which we exist. We come to know we belong and matter through moment-by-moment, attuned, safe, stable, and nurturing relationships in our first moments, early years, and all across life.

These connected, affirming relationships are required for healthy brain development and give rise to the embodied and felt sense of mattering we need to create a life of meaning and purpose and awaken our capacity to nurture ourselves, others, and the natural world that sustains us. Without this, the spark and will to harness our capacity to change, reshape, and heal ourselves, our climate, and our planet are at stake. The loneliness of our hearts in the absence of the belonging and connection we need cuts us off from ourselves, each other, and the natural world; too often leaving us with a sense of emptiness and lack of purpose or meaning. We languish and forget we are built to adapt and grow through adversity. We allow others and nature to languish as well. Awakening to our innate mattering brings us back to life.

Research is clear that our early life positive and adverse experiences shape us - body, mind, and heart. At least two-thirds of adults and half of our children have been exposed to the types of adverse childhood experiences (ACEs) that can steal the sense of safety, belonging, and mattering everyone needs to flourish and give their best. ACEs include emotional or physical neglect and abuse, sexual abuse, family/household dysfunctions like alcohol or drug abuse, poor parental mental health or loss of a parent to suicide, disease or incarceration, as well as living in a toxic environment of racism, discrimination, deprivation, or violence.[1] Fortunately, research also shows that an influx of intentionally created positive childhood experiences (PCEs) and relational healing methods can interrupt the traumatic impact of ACEs and their far-reaching impact on our health, society and, in turn, our environment and climate. Ironically, PCEs most profoundly arise out of how we care for our children and each other when traumatic experiences occur, or when things are difficult or challenging. Feeling safe to talk about feelings, feeling supported when things are hard, being able to safely share about things that really matter when we struggle – these create the sense of mattering that is at the core of PCEs.[2]

The science of mattering and the creation of sustainable flourishing speaks to the deeper need to belong and share in the meeting of adversity and healing for not only ourselves, but also for others. Healing our climate requires we heal and change individually and collectively. In turn, this requires we believe that we matter and can make a difference.

As we awaken to the possibilities and need for safe and nurturing connections and a sense of meaning and mattering, research suggests that we will naturally begin to reset our course toward flourishing and tap into our potential to transform our collective trauma into the inspired collective action we need to heal our hearts, our hope, and our climate. With this as our compass, the changes required to restore and sustain our environment and climate will nurture us in kind, even if what has become familiar falls away, making way for the flourishing of all life and future lives to come. This requires us to reach for the real nourishment that comes from an enduring sense of meaning, purpose, and interconnectedness.

It is a blessing that what we most long for and are biologically wired to seek – belonging and mattering – is also the path forward to simultaneously heal ourselves and our environment. We can restore a shared sense of mattering, heal our collective trauma, and foster an unstoppable drive to seek positive sustainable solutions for both our society and our planet. Changes in mindsets, policies, and practices in our health, social, and educational institutions are already taking hold as we awaken to the promise of a flourishing life for all people and future generations, despite and perhaps even through the challenges we now face. The reality of interconnectedness should give us hope and inspiration. As you will see in the pages ahead, we indeed matter more than we think.

Christina Bethell, PhD
Bloomberg School of Public Health
Johns Hopkins University

INTRODUCTION

This book is for you, and for all people who are interested in social change and open to the possibility that each of us can contribute to a "quantum leap" to sustainability. It's also for people who are concerned about the state of the world and may be feeling a deep anxiety about the future. Above all, it is a book about mattering, especially about how we translate the abstract idea of "mattering" into large-scale systems change that is not only rapid and effective, but also equitable, ethical, and sustainable. To make this translation, we need to take a different perspective and consider new ways of relating to ourselves, each other, the environment, and the future. We need a new way of thinking about social change.

This book explores a perspective that I refer to as quantum social change. It draws on research from the emerging field of quantum social science, which investigates whether and how the concepts, interpretations, and formal models of quantum physics can be used in fields such as international relations, economics, finance, psychology, and sociology.[1] My goal is not to explain quantum social change, but to present a different way of *relating to change*. The nature of the quantum world, which is described by concepts such as entanglement, complementarity, indeterminacy, nonlocality, potentiality, and quantum leaps, invites us to explore what we consider real, and it draws attention to the relationships between mind, meaning, and matter. It gets us thinking about our agency and potential to act in time to make a difference. To really matter.

The question of whether and how we matter is important. We are living in a decisive period in human history, a time when our actions will have profound consequences for life on earth for millennia to come.[2] Thousands of scientific reports and articles have emphasized the risks, impacts, and vulnerabilities associated with different scenarios of climate change.[3] Even without climate change, the degradation of ecosystems and loss of species are diminishing the richness and potential for a thriving world. Both climate change and biodiversity loss have widespread implications for society, particularly in relation to equity, health, well-being, and human security.[4] Still, we are assured by scientists and policymakers that "the future is a choice."[5] But is it really a choice? Can we make a difference?

Coming from the social sciences, I feel provoked when people tell me that it is too late to do anything about climate change, and that we should be preparing ourselves for societal collapse and extinction. I know that the Doomsday Clock is set at "100 seconds from disaster."[6] I have heard convincing arguments as to why it's already too late for many communities and species, and I understand that there are numerous social theories explaining why transformative change is so difficult. However, I also believe that we can do better than merely coping or adapting to crisis upon crisis, including an existential crisis. To do better requires thinking differently, acting differently, and *being* different. By being different, I mean relating differently to ourselves, each other, the environment, and social change.

Fortunately, there is an incredibly large and inspiring literature from the social sciences and humanities suggesting that the promise for social change lies in how we think about it, how we talk about it, and not the least how we relate to it. It also depends on the questions that we ask. As a researcher, I want to understand how we can transform society at the rate, scale, and depth that is called for at this moment in time. Not *whether* we can do it, but *how*. And since we know that not all transformations will lead to fair and just outcomes, I want to understand how we can do so in an equitable, ethical, and sustainable manner. This matters for all of us. How do we shift the cultures and systems that are currently perpetuating the climate crisis? How do we maintain species and ecosystems so that all life can thrive? Are we underestimating our collective capacity for social change?

Climate change is my entry point, as it has been the focus of my research for more than thirty years. In this book I will explore what quantum social science has to offer when it comes to understanding social and human relationships, our relationships with nature and the environment, and our potential to transform all of these. To be honest, I put the manuscript aside many times, doubting that it would ever add anything meaningful to discussions of climate change, and recognizing that using the word quantum with anything but the natural sciences raises eyebrows for being unscientific or "New Age." Of course, many social scientists and humanists would argue that our social reality should not be reduced to *any* type of physics. Period. Some scholars charge that quantum concepts have been misappropriated and misinterpreted to serve questionable interests and agendas.[7] Yet given the nature of global crises, maybe this actually is an appropriate time to consider how meanings, metaphors, and methods informed by quantum physics can inspire social change, and in particular our responses to climate change. In fact, the stakes with climate change are arguably too high *not* to challenge the mindsets of certainty and determinacy that lure us into believing that there is nothing that we can do about it. What we do in this decade really does matter.

In the chapters ahead, I invite you to think about the role of meaning and mattering in a changing climate. We will explore this through paradigms, beliefs, relationships, metaphors, entanglement, consciousness, agency, and fractals. If there is a single take-away message from this book, it is that you matter more than you think. Literally. This is because individual change *is* collective change; we are inherently entangled through shared language, meaning, and values. Our relationships to concepts and ideas such as *self*, *other*, *nature*, and *change* have been influenced by classical Newtonian physics and its assumption of separateness. Quantum physics and quantum social science challenge this assumption by recognizing that entangled quantum systems are never fully separable.[8] This is not to say that individuals are completely inseparable, or that identities and boundaries are unimportant. However, from the perspective of quantum physics, connections, relationships, and communication are foundational, and this has implications for social change.

Quantum social change describes a conscious, nonlinear, and non-local approach to transformations that is grounded in our inherent oneness. It recognizes that we are entangled through language, meaning, and shared contexts, and that our deepest values and intentions are potential sources of individual change, collective change, and systems change. This recognition, when expressed through a particular quality of agency, can shift systems and cultures in a manner that is both equitable and sustainable. This moves us beyond a classical and mechanistic search *for* leverage points to transform systems. Instead, it suggests that *we* are the leverage points, and that how we show up in every moment matters. Quantum social change does not require us to wait for some remarkable leader or hero to introduce solutions that will save us – it is about each of us acting right now, within our own dynamic context and spheres of influence, to generate new patterns and relationships.

The idea that we each have the potential to transform systems at a global scale involves a subtle yet profound shift, and words alone cannot achieve this. The assumptions of a classical paradigm have long been questioned, not the least by poets, writers, musicians, and artists who have found creative ways to express the limitations of the dominant Western worldview. The artwork in this book, created by Tone Bjordam, thus offers another way of expressing our entanglement with nature. Reflection questions at the end of each chapter are included to trigger insights and ideas that may help you to consider how to integrate and embody a quantum perspective on social change. My hope is that this exploration of quantum social change will inspire and empower you to engage consciously, creatively, collaboratively, and not the least, courageously with the transformations needed to realize an equitable and thriving world for all.

1

THIS DECADE

THIS DECADE MATTERS. We are living at a time when the existential threats associated with climate change are challenging us to transform ourselves, our systems, and our societies. Climate change, interacting with many other global trends, is diminishing the potential for all life to thrive. Scientists tell us we have about a decade to "bend the curves" for greenhouse gas emissions if we are to avoid dangerous climate change. We have been informed of the risks and dangers facing us if we fail. The situation is serious, and if we take the science seriously, we need to transform at a rate and scale that is unprecedented in human history. Do we have time to change? More important, can we consciously change? Is it possible to create a world where everyone can thrive? If so, what is our role in making this happen? Do we matter?

Profound questions arise during times of crisis, and deep down, we all have ideas about what matters. For many people, what matters is being alive, feeling connected, and sensing that we are significant and that our lives are meaningful. We share an innate longing for equity, dignity, compassion, and love. Science writer John Horgan describes humanity as "matter that yearns to matter."[1] If there has ever been a moment in history when a deep inquiry on mattering is relevant, it is now.

ON MATTERING

Mattering means being important or significant. It also refers to a physical substance, as distinct from mind and spirit. Current understandings of matter draw heavily on the mechanistic worldview that is associated with the 18th Century Enlightenment, also referred to as the "Age of Reason." In the Western world, this era was characterized by the philosophical search for truth through knowledge, reason, and experimentation. Newton's laws of motion and gravitation, along with the four laws of thermodynamics, contributed to new understandings of how the world worked. Simply put, the world was seen as a machine made up of separate parts that function together. Matter, defined as a substantial "thing" located in or moving through space and time, could be described down to the smallest level. In other words, matter was something that could be studied, understood, and manipulated.

This modern and Western development of science has brought important benefits, such as a longer life expectancy and an improved quality of life for many people. Moreover, classical physics has contributed to countless scientific breakthroughs and discoveries, including steam engines, trains, automobiles, and airplanes. Powered by the energy of fossil fuels, these inventions have accelerated the release of stored carbon into the atmosphere, changing its composition and influencing the global energy balance, which has had profound consequences for the climate system, ecosystems, biodiversity, and human health and wellbeing.

The influence of a mechanistic approach is evident in climate change research, which uses mathematical models to understand complex interactions among different parts of the Earth System, including the atmosphere, biosphere, hydrosphere, and cryosphere (ice cover). These models, together with insights from complexity science, have made important contributions to understandings of how human activities are influencing the global climate system, as well as how humans will be affected by the consequences. Take, for example, the "wiring diagrams" of Earth Systems science. These conceptual figures use boxes and arrows to depict key relationships within the Earth System, as shown in artist Tone Bjordam's drawing of the well-known "Bretherton Diagram" (Figure 1). Although this diagram recognizes that human activities influence the Earth System and create impacts, humans themselves seem to be conceptually external to the system. In particular, the potential for humanity to reflect, imagine, collaborate, and consciously co-create equitable transformations has not been adequately integrated in sustainability science.[2]

TIME MATTERS

Earth Systems Science provides us with valuable insights on past, present, and future trajectories of global processes, including how they have been influenced by humans. This knowledge has led scientists to propose a new geological epoch referred to as the Anthropocene, a term that acknowledges that humans have become a powerful force in shaping the future of the planet. However, when we focus on geologic time scales, the possibility for us to change course feels very limited. In fact, when we look at current and projected trends in carbon emissions from the production and use of fossil fuels relative to the UN's Sustainable Development Goals (SDGs) and the Paris Agreement, it is easy to conclude that the potential for change is indeed small. This is reinforced by understandings of the momentum and inertia in both climate and energy systems: the storage of a substantial amount of heat in the oceans and the lifetime of carbon dioxide in the atmosphere make it clear that warming will continue for decades to come, even with successful mitigation efforts. Add to this our understandings of planetary boundaries, thresholds, and tipping points in Earth System processes, together with the seemingly slow pace of social change and the lack of real political engagement, and it does seem hard to see how you and I can make a difference.

This represents a paradox – one of those self-contradictory statements that at first seems true. Most people recognize that humans matter when it comes to transforming the global environment;

Deep Sea Sediment Co

Land and Ice

Mapping

Continents & Topography

Atmospheric Physics/Dynamics

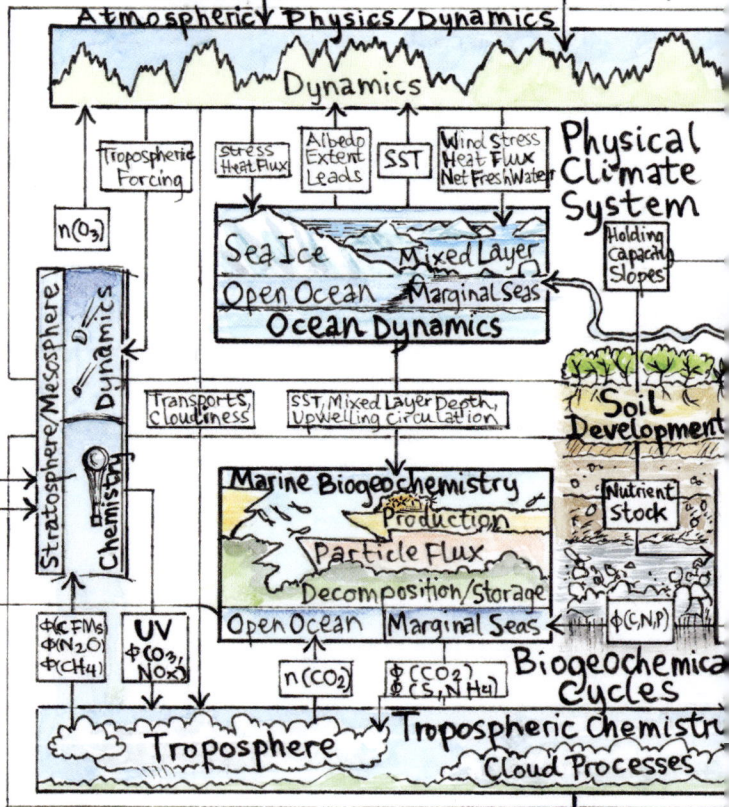

Dynamics

| Tropospheric Forcing | Stress Heat Flux | Albedo Extent Leads | SST | Wind Stress Heat Flux Net FreshWater |

n(O₃)

Physical Climate System

Holding Capacity Slopes

Sea Ice
Open Ocean
Mixed Layer
Marginal Seas
Ocean Dynamics

Volcanism

Φ(H₂O)
Φ(S,N,...)

UV, Particles

Solar/Space Plasmas

Stratosphere/Mesosphere Dynamics
Chemistry

Transports Cloudiness

SST, Mixed Layer Depth, Upwelling Circulation

Soil Development

Nutrient Stock

Marine Biogeochemistry
Production
Particle Flux
Decomposition/Storage
Open Ocean
Marginal Seas

Biogeochemical Cycles

Φ(CFMs)
Φ(N₂O)
Φ(CH₄)

UV
Φ(O₃, NOₓ)

n(CO₂)

Φ(CO₂)
Φ(S,N,Hal)

Φ(C,N,P)

Troposphere

Tropospheric Chemistry
Cloud Processes

Foraminifera (Temperature)

Deep Sea Sediment Cores

n(CO₂)

Ice Cores

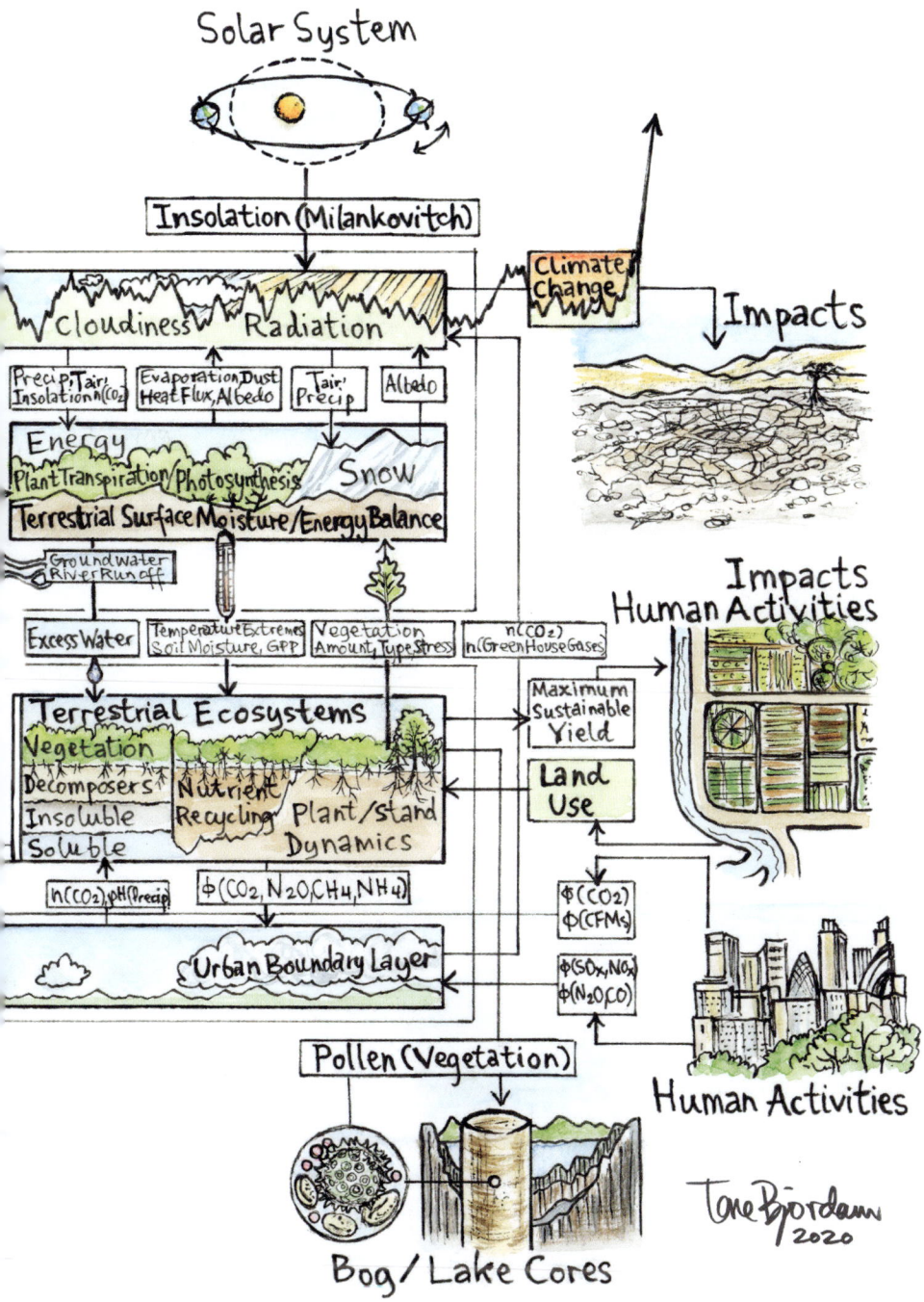

Solar System

Insolation (Milankovitch)

Climate Change

Impacts

Cloudiness Radiation

Precip, Tair, Insolation n(CO₂)

Evaporation, Dust Heat Flux, Albedo

Tair, Precip

Albedo

Energy

Plant Transpiration/Photosynthesis

Snow

Terrestrial Surface Moisture/Energy Balance

Groundwater River Runoff

Excess Water

Temperature Extremes Soil Moisture, GPP

Vegetation Amount, Type, Stress

n(CO₂) n(GreenHouse Gases)

Impacts Human Activities

Maximum Sustainable Yield

Terrestrial Ecosystems

Vegetation

Decomposers

Insoluble

Soluble

Nutrient Recycling

Plant/Stand Dynamics

Land Use

n(CO₂), pH (Precip)

φ(CO₂, N₂O, CH₄, NH₄)

φ(CO₂) φ(CFMs)

Urban Boundary Layer

φ(SOₓ, NOₓ) φ(N₂O, CO)

Pollen (Vegetation)

Human Activities

Tone Bjordann 2020

Bog/Lake Cores

Figure 1: Earth System "wiring diagram" (based on Bretherton Diagram).

this is the core message behind the idea of the Anthropocene. However, many of us feel that we do not matter when it comes to deliberately transforming ourselves, our societies, and our systems to meet the challenges that we are facing today. Let's look closer at this "paradox of mattering."

On the one hand, we are told that the solutions to climate change are straightforward. In fact, the Intergovernmental Panel on Climate Change (IPCC) emphasizes that we already have most of the technologies we need to dramatically reduce greenhouse gas emissions, and countless scientific papers and reports describe blueprints and road maps for transitioning to low-carbon societies.[3] Scientists remind us that the sun, wind, water, waves, friction, and the Earth itself are sources of heat and motion that can be used to produce renewable energy and electricity. Forests, vegetation, and soils take up or "sequester" carbon from the atmosphere and store it in the biosphere. Environmental groups continually inform us of the dozens of things we can do to address climate change. This includes everything from buying energy-efficient appliances and reducing meat consumption to taking actions to influence climate politics.[4] Project Drawdown is monitoring 100 existing and promising solutions that can arguably *reverse* global warming and contribute to the 17 SDGs, and many communities around the world are now engaging with these solutions.[5]

On the other hand, we also know that greenhouse gas emissions have increased over the past decade, and that energy use from fossil fuels continues to rise faster than the alternatives.[6] When we do take environmentally-friendly actions, it can be discouraging to read that 100 fossil fuel companies have been responsible for 71% of greenhouse gas emissions since 1988, that banks and oil companies are continuing to invest in fossil fuels, and that the contributions to global CO_2 emissions from aviation and transportation have been growing with the expansion of the travel industry.[7] Even our digital lives take up more and more energy, with cloud storage and internet streaming accounting for an increasing percentage of greenhouse gas emissions.[8] The problems are systemic, and we are far off track from the goal of limiting global warming to below 2°C. Furthermore, we have done little to reduce the risks and vulnerabilities associated with the impacts of climate change. We are not adapting well to a climate that we are changing.

Greenhouse gas emissions patterns are highly skewed, with ample research showing that those with the least responsibility for climate change are the most vulnerable to its impacts, and that marginalized communities and social groups often have little capacity to adapt to changes in climate variability and extreme events.[9] Exposure to multiple stressors, including those associated with economic and financial globalization, exacerbate existing inequalities, affecting livelihoods and lives across generations.[10] At a time when a growing number of people are demanding that politicians take environmental justice seriously, political debates are becoming increasingly polarized and contentious. Some countries are moving backwards rather than forwards in terms of energy standards, pollution controls, and deforestation policies. The actions we take as individuals seem to have little influence on the dynamics of power and interests, and on the transformations of cultures and systems that are needed to address the climate crisis.[11]

A sense of urgency calls for radical change, where radical means getting to the root of the problem. Crises always introduce opportunities to think and act differently, and to address the root causes of problems. They can also open up new ways of approaching relationships, including our relationship to change. The ways that we think about social change have been generally informed by research on social norms, social practices, social movements, collective action, and power dynamics, each reflecting different ways of thinking about politics, relationships, and causality. Yet these relationships are typically based on classical assumptions about the nature of the social world, particularly the idea that we are inherently separate from each other and from nature.[12] As anthropologist Arturo Escobar writes, "the questions of wholes, form, and coherence remains unsolved in social theories."[13] These assumptions have implications for the politics of social change, including understandings of political agency.[14] Fortunately, we have the opportunity to break with the assumptions of the past and engage with new ways of thinking, being, and doing.

TRANSFORMATIVE CHANGE

Change is at the heart of much of social science research. More recently, the concepts of transformation and transformative change have been used to express the type of change that is needed to respond equitably and sustainably to multiple global crises.[15] Transformation can be defined as significant changes in form, structure, or meaning-making. It can also be described as a much deeper process, including "the unleashing of human potential to commit, care, and affect change for a better life."[16]

Inherent in most definitions of transformation are implicit and often unquestioned theories, understandings, and assumptions about causality, particularly regarding the relationship between individual change, collective change, and systems change. For example, a classical, materialist depiction of development has led to the pursuit of progress, innovation, and growth for their own sake, rather than for the well-being of people and the planet. Again, this is not to deny the significance of the Enlightenment worldview, but it is essential to recognize its limitations and negative consequences for many people and much of the planet. This depiction also downplays the significance of free will and personal responsibility – or "response-ability." A focus on the purely functional connections between the different parts of the machinery that drive our systems ignores deeper questions of meaning, purpose, and what *really* matters.

Over the past 150 years, this mechanistic worldview has been challenged by researchers in the social sciences, humanities, philosophy, theology, and even the natural sciences. This includes the idea that matter is lifeless and dead, that consciousness is an illusion created by the brain, and that individuals are separate and distinct from each other and nature. Of course, many religions, wisdom traditions, and cultures have long recognized these ideas, including the interplay between particularity and universality, i.e., the dynamic relationship between parts and wholes. For example, African Ubuntu philosophy is based on the understanding that "I am because we are," and many cultures view landscapes as relational, alive, and with agency.[17] Native American and Eastern philosophies also describe the holistic nature of the universe.[18] Holism refers to the philosophical idea of an interconnectedness that can only be explained with reference to the whole, rather than to its parts. From a holistic perspective, conceptual distinctions and dualities such as "mind and body" and "humans and nature" are not separate but one.

Building on these age-old ideas, a school of thought within academia referred to as "new materialism" considers matter in a relational, emergent sense.[19] Feminist philosophers and social theorists have challenged the colonization of people, cultures, and ideas based on a Western worldview. For example, political theorist Jane Bennett talks about matter as vibrant and alive, something that opens up for unexpected possibilities that are emergent, rather than linear and deterministic.[20] From the perspective of new materialism, all matter – including non-human beings, things, or entities – has agency and the affective capacity to influence the world. Again, this is not a new idea. As Jerry Lee Rosiek and colleagues point out, it was developed by indigenous thinkers and scholars thousands of years before contemporary philosophers of science "discovered" it.[21]

The idea that both humans *and* non-humans are vibrant and have agency has been referred to by biologist and philosopher Andreas Weber as an "Enlivenment" worldview.[22] In contrast to the Enlightenment worldview, this perspective stresses the importance of incorporating lived experience and embodied meaning into our understanding of the biosphere. It recognizes the biosphere as connected, vibrant, and alive – a view that can potentially transform the way we relate to nature, each other, and the future. As Weber describes it, an Enlivenment worldview does not seek to replace the science of the Enlightenment, but rather to expand it by challenging some of its core assumptions.[23]

The core assumptions of the Enlightenment worldview have been questioned by physicists themselves, particularly those working in the realm of quantum physics. Quantum physics, which refers here to both quantum mechanics and quantum field theory, challenges classical models of reality at the subatomic scale. Electrons can be in two places at once and tunnel between physical barriers? Particles are nonlocally entangled? Time goes backwards? Although these are intriguing ideas, a quantum lens has not been taken seriously when it comes to understandings of social change. Why not? Many would argue quite strongly that it should not, since quantum physics applies to subatomic, atomic, and molecular scales. Yet as physicist Christopher Fuchs argues, "What is at stake with quantum theory is the very nature of reality. Should reality be understood as something completely impervious to our interventions, or should it be viewed as something responsive to the very existence of human beings?"[24] A quantum lens, with all of its ambiguity, may provide a fresh perspective on social change.

In short...

We are living in the decade that matters, and we have a window of opportunity for addressing climate change, biodiversity loss and other sustainability challenges. Yet at a time when many people long to be considered significant, we face a "paradox of mattering." More and more people recognize that we are changing the global environment, yet we still struggle when it comes to consciously transforming ourselves, our systems, and our societies in an equitable and sustainable manner. Responding to climate change with clarity, coherence, and most of all, wisdom, requires a strategy that transcends the urgency of the moment and compels us to engage with new ways of thinking, being, and doing. Now is a good time to broaden our understanding of "mattering" by exploring new paradigms for social change.

—— Reflections ——

- Do you think the next ten years are important when it comes to climate change responses and deep social transformations? Why or why not?

- What matters most to you? What kind of life would you like for yourself and others in ten years?

- How do you as an individual feel that you can best contribute to the transformations that are called for at this time?

2

PARADIGMS

PARADIGMS MATTER. They represent the dominant thought patterns that underlie theories and methods of science, as well as policies and practices related to how we organize society. Paradigms influence the way that problems are defined and addressed, including what is considered realistic, legitimate, and effective. They shape the ideas, concepts, images, metaphors, and memes that we use to describe reality. Paradigms are considered by environmental scientist Donella Meadows to be a powerful leverage point for systems change, i.e., a place where a solution can be applied such that it results in a systems shift or transformation.[1] Systems, broadly defined as a set of things that work together to form a whole, are made up of relationships between different components. A change in how these relationships are conceived, understood, and approached can have a profound influence on the way systems evolve.

Climate change offers an opportunity to take a closer look at how paradigms shape our approaches to not only systems change, but social change. Given the impacts of climate change that have already been experienced, the uneven distribution of losses and damages, and our understanding that the risks will be exponentially higher as greenhouse gas emissions continue to rise, their power and potential should not be underestimated. Indeed, philosopher Thomas Kuhn emphasizes that "probably the single most prevalent claim advanced by proponents of a new paradigm is that they can solve problems that have led the old one to a crisis."[2]

QUANTUM THEORY

Over a century ago, discoveries in quantum physics introduced a new and ambiguous understanding of matter at the subatomic scale, which challenged the classical scientific paradigm that has influenced modern Western thinking.[3] As physicist Henry Stapp explains, quantum mechanics replaced Isaac Newton's description of matter as "solid, massy, hard, impenetrable, moveable particles" and swept away the meaningless "billiard-ball" universe.[4] This analogy compares matter in the universe to balls rolling on a billiard table, with their direction causally-determined by previous events. Reflecting on 100 years of quantum physics, Daniel Kleppner and Roman Jackiw emphasize that "Quantum mechanics forced physicists to reshape their ideas of reality, to rethink the nature of things at the deepest level, and to revise their concepts of position and speed, as well as their notions of cause and effect."[5]

In contrast to a mechanistic and deterministic world, quantum physics describes a universe of indeterminacy, entanglement, superposition, complementarity, uncertainty, and not the least, potentiality.[6] Significantly, it shows that at the subatomic level, electrons and photons of light can behave as both particles and waves. However, although they have both attributes at the same time, the principles of complementarity and uncertainty tell us that we cannot observe or measure both wave- and particle-related properties at the same time. This is true for other properties as well, such as position and momentum: the more we know about the momentum of a particle, the less we know about its position, and vice versa. Yet quantum indeterminacy is not just about the limits to knowledge. It is about the very nature of reality itself, in that matter at the subatomic scale seems to be nothing more than an ambiguous cloud of probability until a measurement or observation is made.

Quantum coherence is another feature of the quantum world, whereby the wave functions of two or more particles are in a state of superposition that can be described with one equation. Superposition means that particles can be "entangled" – in other words, non-locally correlated to predictably and instantaneously interact with each other, regardless of the distance between them. Measuring one entangled particle provides information about others, at least in closed systems where information is not lost to the surrounding environment. The notion of a wave-particle duality and the concepts of quantum coherence and entanglement have raised many metaphysical questions about our foundational understandings of reality, which are reflected in diverse interpretations of quantum physics.

Still, the metaphysical implications and deeper significance of quantum physics have been considered less important than the practical functionality of the mathematical equations. From a mathematical perspective, quantum mechanics stands out today as the most successful theory in physics, and an instrumental, "shut up an calculate" approach has been used to develop computer chips, cell phones, the internet, and many other technologies that we use today. [7] These technologies have undoubtedly transformed the social reality that we are experiencing today, including how we connect with others.

Upcoming breakthroughs in quantum computing and quantum encryption are likely to transform society even further in the 21st century and beyond, and this is sure to have implications for the global environment, not to mention geopolitics. For this reason, it is important to consider the deeper implications of quantum physics for understandings of social change, even if only in the form of a thought experiment or speculative inquiry. Importantly, many theoretical developments in quantum physics start out as thought experiments, and only later are they corroborated or falsified through experimental research. For example, empirical results now confirm that outcomes are dependent upon the observer, which was once only speculation.[8] Lab research also shows that objective reality does not exist, at least for local models of quantum systems.[9] While the concept of entanglement has been theoretically restricted to subatomic particles, recent experiments indicate that molecules made up of more than 2000 atoms (which is the size of some proteins) can exist in a wave-like state.[10] This research is important, and it seems to support physicist Carlo Rovelli's assertion that "reality is not what it seems."[11]

INTERPRETING REALITY

Many interpretations of quantum physics have been presented in academic journals and popular science books, making it clear that there is no consensus on the foundations of our physical reality. Among the most discussed and debated interpretations of quantum physics is Niels Bohr's Copenhagen School, which includes a variety of perspectives based on the idea that a probability wave "collapses" into a reality when a measurement is made. This is a radical proposition, particularly from a classical, Newtonian perspective, yet it has been confirmed through experiments. For example, the famous double-slit experiment carried out by physicists in the 20th Century demonstrates that when electrons or photons are shot through a small slit, they can behave as both classically-defined particles *and* waves, depending on whether the measurement was observed.

The idea that measurements are involved in the collapse of the wave function implied to Bohr that matter is not separate and distinct from the observer. In *Meeting the Universe Halfway: Quantum Physics and the Entanglement of Matter and Meaning*, physicist and feminist theorist Karen Barad notes that this points to a key lesson from quantum physics: *"we are part of the nature that we seek to understand."*[12] From a non-dual perspective, observations are better understood as interventions, where outcomes are influenced by the instrument, device, or sensor that performs the observation or measurement. Humans, as sensory beings, have been regarded by some as instruments that can influence measurements, and by others as an entangled part of the phenomenon that is being measured.[13] From this perspective, we are not distinct from the systems that we are measuring – we are the systems. This represents a significant contrast to the "view from nowhere" of an objective scientist, which has been the foundation of classical science.[14]

Not surprisingly, numerous alternative interpretations of quantum mechanics have been proposed to challenge the Copenhagen interpretation. For example, the "Many-Worlds Interpretation" resolves the question of the wave function collapse in a mathematically elegant manner by considering our universe to be just one among an infinite number of universes. Thus, when a wave function collapses, each possible outcome is enacted in one of these parallel "worlds." In other words, every quantum event branches off into a different reality, within some other world.[15] According to this interpretation, our world is the way it is, simply because it represents only one of an infinite number of possibilities.[16]

Some interpretations of quantum physics directly address the role of subjectivity and consciousness in shaping reality. For example, an interpretation based on quantum information theory, referred to as QBism (short for Quantum Bayesianism), draws on personalist Bayesian probability and suggests that a quantum state represents the degree of subjective beliefs about possible outcomes that an individual holds. The quantum wave function then constantly changes as agents update their beliefs about the system.[17] This interpretation acknowledges that "reality is more than any third-person perspective can capture," and introduces the concept of participatory realism, i.e. that humans are an integral part of a quantum reality.[18] In recognizing the significance of personal experience, QBism brings the subject back into science.[19]

Another example of how quantum physics brings the subject back into science is Henry Stapp's interpretation of the quantum Zeno effect, a phenomenon in physics where repeated observations of a radioactive particle can prevent it from decaying in the usual, predicted manner. Stapp extends this to argue that the deliberate application of mental effort or intention holds in place our brain's "template for action," which then produces the brain states that subsequently generate experiential feedback.[20] As a result, Stapp contends that we live in "a universe in which we human beings, by means of our value-based intentional efforts, can make a difference first in our own behaviors, then in the social matrix in which we are imbedded, and eventually in the entire physical reality that sustains our streams of conscious experiences."[21] This theory presents a new understanding of ourselves and our place in nature, and raises important sociological and philosophical issues that, according to Stapp, "extend far beyond the narrowly construed boundaries of science."[22]

Most scientists consider quantum physics to be interesting and significant either at the scale of subatomic particles, i.e., electrons, protons, neutrons, quarks, etc., or at the cosmic scale, including black holes. However, in a macro world that is warm, wet, and full of noise, quantum coherence and its effects wash out and hence are considered irrelevant. Aside from the impacts of the many technologies developed based on quantum physics, the macroscale world of our daily social lives is considered to operate exclusively within the domain of classical, Newtonian physics, with its foundations in the Cartesian dualism of mind and matter. There is an implicit assumption that only the deterministic laws of classical physics influence our social world and the construction of social life.[23]

A classical understanding of "human nature," and "human-environment relationships" leads many scientists to feel quite certain that we, as a species, are imminently doomed by climate change.

For example, in a bestselling popular science book called *Seven Brief Lessons on Physics*, Carlo Rovelli concludes his explanation of modern scientific theories by reminding us that humans are only a small part of the universe, and that we are subject to the norms of nature. He ends with a profoundly pessimistic conclusion about the future of humanity:

> I believe our species will not last long. ... We belong to a short-lived genus of species. All of our cousins are already extinct. What's more, we do damage. The brutal climate and environmental changes which we have triggered are unlikely to spare us. For the Earth they may turn out to be a small irrelevant blip, but I do not think that we will outlast them unscathed – especially since public and political opinion prefers to ignore the dangers which we are running, hiding our heads in the sand.[24]

Rovelli is not the only scientist to reach this conclusion. James Lovelock, who developed the Gaia Hypothesis describing how living organisms co-evolve with their environment, once said in an interview with *The Guardian*, "I don't think we're yet evolved to the point where we're clever enough to handle as complex a situation as climate change."[25] In a world where individuals are perceived to be intrinsically separate and distinct from one another and from nature, and in a world where consciousness and free will are considered illusions, it is easy for experts to look at today's situation and conclude that humans simply do not have the capacity to successfully respond to the challenge of climate change. These ideas also penetrate our culture. For example, as novelist Jonathan Franzen writes in *The New Yorker*, "the climate apocalypse is coming. To prepare for it, we need to admit that we can't prevent it."[26] Such statements suggest that regardless of what we would like to think or do, it's already "game over." We do not really matter.

REALITY CHECK

Interesting. Scientists tell us that the universe of planets, stars, gravity, space, time, and matter at the subatomic level is characterized by uncertainty, indeterminacy, and mystery. Yet the role of humans in this universe is considered to be certain and predictable. In fact, applying a quantum lens to the macroscale and invoking consciousness tends to be ridiculed as belonging in the category of "woo-woo" science, particularly when it is used in explanations of the mind or brain functioning.[27] Such ridicule can be effective in limiting the scope of scientific inquiry into understandings of social change. What is important here is not to reduce all ways of knowing to physics, but to open up the inquiry to consider how subjectivity influences our experiences of the social world. In fact, thought experiments, imagination, and speculation have often been potent drivers of change, and it is worth reflecting on the risks of maintaining classical assumptions of the social world at any cost.[28]

In *How the Hippies Saved Physics: Science, Counterculture, and the Quantum Revival*, David Kaiser describes the freewheeling speculation and philosophical engagement with physics and consciousness among members of the Fundamental Fysiks Group in California in the 1970s and 1980s.[29] Their work drew attention to "Bell's theorem" and its implications for notions of nonlocality and entanglement. This theorem, developed by John Bell, rules out the possibility of hidden variables to explain quantum physics, thereby confirming entanglement and nonlocality. The group also explored the relationship between quantum physics and Eastern philosophies, pointing out that the idea of complementarity can be traced back 2,500 years to Chinese philosophy and its emphasis on the dynamic interplay between yin and yang.[30] Their inquiry helped to instigate major breakthroughs in quantum physics, including quantum encryption. If novel thought experiments such as these can lead to scientific breakthroughs, there may also be value in speculating about alternative ways of approaching social change.

Social scientists have been increasingly exploring the macro-scale social implications of quantum technologies and quantum theory, including the implications for peace and security.[31] The core mathematical principles of quantum theory are already being used to better understand and describe the complexity of human behavior and decision making.[32] For example, quantum decision-making recognizes that judgments and decisions are influenced by context, and it allows for preference reversals. As cognitive scientists Jerome Busemeyer and Peter Bruza argue, "Quantum theory provides a fundamentally different approach to logic, reasoning, probabilistic inference, and dynamic systems."[33] Quantum social science is "*not* about reformulating social science on a quantum size scale," but rather about investigating social problems with the help of formal models and concepts used in quantum physics.[34] Going deeper, it offers a lens for re-thinking human behavior, and for exploring the philosophical implications of quantum theory for our shared social reality.[35]

The point here is *not* that we should simply choose an alternative paradigm based on a particular interpretation of quantum physics. There is a danger in swapping one set of beliefs with another, without critical reflection or evidence to support it. In fact, there is good reason to be critical of alternative paradigms, as psychological and emotional appeals have sometimes been strategically used by religious institutions, sects, cults, political parties, governments, and scientists to promote thoughts that support particular agendas. We can, however, critically engage with alternative paradigms, and maintain an open mind about what the latest science is telling us about the nature of relationships, including our relationship to nature.

It is important to emphasize that classical and quantum physics do not actually correspond to two separate domains of reality. In fact, classical behavior can be considered an emergent property of a quantum system. In other words, the quantum versus classical distinction is a false dichotomy.[36] This is evident in the field of quantum biology, where quantum processes have been identified as playing an important role in photosynthesis, sense of smell, and bird navigation. These discoveries show that "relatively small numbers of highly-ordered particles, such as those inside a gene or the avian compass, can make a difference to an entire organism."[37] If the dividing line between the micro and macro world is blurred, maybe the dividing lines between "us" and "others" and "humans" and "nature" are also ambiguous.

IN SHORT...

Paradigms have power, and they influence how we describe reality. The paradigm shift from classical to quantum physics has transformed our understanding of reality, introducing notions such as uncertainty, potentiality, the wave-particle duality, and quantum entanglement. It shows us that reality is not what it seems: individuals may not be separate and distinct from one another and from nature after all, and consciousness and free will may be more influential than we believe. If both humans and non-humans are participants in an ambiguous quantum reality, we might want to revisit some of our beliefs about the social world. In fact, what if *social* reality is not what it seems?

——— REFLECTIONS ———

- What do you think quantum physics tells us about the nature of reality?

- Do you think that humans are capable of addressing complex problems such as climate change by creating an equitable and thriving world? Why or why not?

- Have you ever questioned a dominant thought pattern or paradigm? What happened – or what might happen if you did?

3

BELIEFS

BELIEFS MATTER. They shape the reality that we experience, including our shared social reality. But what exactly are beliefs? One common definition is an acceptance that something is true or exists, even when there is no direct evidence or proof. Our beliefs can be influenced by subjective experiences, social and cultural norms, scientific experiments, education, religious or spiritual teachings, or indoctrination. Though it is possible to explore and test beliefs, we are often habituated or socially and culturally conditioned to hold certain beliefs without questioning them. In *Reality is Not What it Seems: The Journey to Quantum Gravity*, Carlo Rovelli points to the importance of challenging our beliefs, recognizing that they may turn out to be wrong, or at least naïve: "The search for knowledge is not nourished by certainty: it is nourished by a radical distrust in certainty."[1]

Our beliefs influence the perspectives we take on the relationship between self and society, including how we perceive a situation, which actions are considered appropriate, and what the effects or consequences might be.[2] Both consciously and unconsciously held beliefs about climate change often reflect other beliefs about human-environment relationships, causality, free will, and mattering. In fact, they play a critical role in shaping whether and how we engage with the issue. In *Mindmade Politics: The Cognitive Roots of International Climate Governance*, Manjana Milkoreit reminds us that belief systems are cognitive structures, and each is a reflection and interpretation of reality that influences our expectation of future developments. As such, beliefs can both "enable and constrain political thought and decision making."[3] Consequently, it is important to pay attention to what we believe and why.

THE POWER OF BELIEFS

"I don't believe in climate change." These six words represent a powerful statement that has profound implications for our collective future. Such a belief typically implies skepticism or antagonism towards both the science of climate change and the policies and initiatives aimed at addressing it, whether through investments in renewable energy, reforestation programs, the development of public transportation infrastructure, political and economic changes, social policies to reduce vulnerability, or an education curriculum that prioritizes transformative learning and critical thinking. This may relate to an underlying belief that we are too small and insignificant to influence the global environment, or that while individuals may indeed create their personal reality, they do not have any responsibility for a shared or collective reality.

The network of beliefs associated with the proposition that the planet *is* warming is based on research using observed and experimental data, climate models, international assessments, and expert opinion.[4] Yet whether or not we believe that humans are responsible for climate change depends not only on scientific evidence, but also on cognitive biases, personal interests, cultural norms, and how we view our relationship to nature.[5] Although beliefs should be tentative and changeable, they are often held as absolute truths, or sometimes as malleable truths that suit personal whims (e.g., "post-truth"). However, there is a distinction between beliefs based on our best understanding of a situation or relationship, and beliefs based on hope, wishful thinking, intuition, guessing, or political expediency. As a result, it is important to back up intuitive beliefs with critical thinking. This is especially important if the beliefs that we accept or dismiss can have life-altering consequences at a planetary scale.

"I don't believe I matter." This is an equally powerful statement, with equally profound implications for our collective future. This belief implies a lack of perceived agency or capacity to influence the external world, or at least at a scale relevant to the challenge of climate change. It also implies an acceptance of current political arrangements, interests, and distributions of power as fixed. Given the data on environmental and social trends, it is easy to understand such feelings of insignificance. In can be heartbreaking to experience what is happening around the world, whether it is record-breaking temperatures, the melting of the Greenland ice sheet and glaciers, the bleaching of coral reefs, or the loss of biodiversity. It can also be difficult to take in the extent of the losses – whether one is measuring them in terms of human lives, species extinctions, or foregone opportunities. We know that more losses will come, and that we have to both cope with and adapt to change. This has contributed to a growing sense of despondency and doom, reinforced by apocalyptic stories and assertions that "it's too late to do anything."

The perception that we do not matter has important consequences for our individual and collective responses to climate change. In a widely-read paper titled "Deep Adaptation: A Map for Navigating Climate Tragedy," Jem Bendell interprets current climate change science as "indicating inevitable collapse, probable catastrophe and possible extinction."[6] This sobering claim about inevitability, probability, and possibility is based on an assessment of the climate change literature and current emissions trends, as well as the lack of an adequate response to the climate crisis. Yet these conclusions seem to be associated with a classical interpretation of social change, where individuals are considered discrete and separate from each other and from nature. Would our probability of survival be greater if we believed that individuals and groups were conscious and entangled agents who are capable of transforming cultures and systems at scale? Without reflecting on what we believe and why, we are likely to continue in default modes, perpetuating the behaviors, structures, and systems that are degrading the environment and contributing to climate change, biodiversity loss, and growing inequality.

PROBABILITIES OF CHANGE

When it comes to equity and sustainability, the belief that all humans and all species matter can be an important motivation for social transformations. Yet when it comes to change, our beliefs about probability and uncertainty are just as important, as they influence how much potential impact we perceive ourselves to have, both individually and collectively. In a mathematical sense, probability describes the extent to which an event is likely to occur. It can be considered a measure of uncertainty. Classical probabilities are typically defined as the ratio of the favorable cases to the whole number of cases possible. However, this "favorable cases divided by all cases" or *frequentist* approach to probability applies to multiple trials, rather than to single cases or individual events. This can be thought of as the difference between rolling a pair of dice numerous times versus experiencing a large piece of the Ross Ice Shelf in Antarctica breaking off and, in the process, indirectly contributing to several meters of sea level rise.

What role do humans play in these cases? With a frequentist approach, the probability lies in the action and the object, rather than in the agent, i.e., the person carrying out the action.[7] In other words, the probability of tossing two sixes when rolling dice is associated with the toss and the dice, rather than the person who is tossing them. Extrapolating this to climate change actions, it is as if the probability associated with choosing a plant-based diet is inherent in the food or act of eating, rather than in the person who is deciding what to eat. A frequentist approach associates the probability of choosing air travel with the airplane and the flight, rather than with the traveler. Probabilities regarding investments in fossil fuels are considered inherent to oil or gas, rather than to the beliefs and values of the people and organizations that are making investment decisions. In short, a frequentist approach does not assess or take account of probabilities and uncertainties related to the people who are taking the actions. In doing so, it ignores their agency and capacity to influence outcomes.

Single-case probabilities are not the same as frequentist probabilities. Referring back to Jem Bendell's perspective on deep adaptation, the probability (or likelihood) that we are facing inevitable societal collapse because of climate change cannot be calculated as favorable cases divided by all cases, as we have only one planet – a single case.[8] Frequentist approaches, which rely on historical observations, are considered to have little value for projecting future trajectories of climate change.[9] How then do we assess the probability of a sustainable planet?

One approach involves mathematical simulation models that have been designed to generate probabilities of future collapse. The World3 computer model developed in the early 1970s by the Club of Rome is one example, and it concluded that if growth trends were to continue, the limits to exponential growth would lead to "sudden and uncontrollable decline in both population and industrial capacity."[10] This model, which has been the subject of extensive debate over the past decades, was based on a systemic analysis of five variables: population, food production, industrialization, pollution, and consumption of nonrenewable natural resources.[11] However, feedbacks related to individual and collective agency or cultural shifts in beliefs, values, worldviews, and paradigms were not modeled. For example, the growth of the animal rights movement and veganism reflects a shift in beliefs, values, and worldviews, which could at some point have a dramatic, nonlinear impact on food production and greenhouse gas emissions.

Another approach, known as agent-based modeling, integrates the behavior of individuals into modeled outcomes. It considers individuals as acting based on simple rules that can generate complex behaviors. Here, humans tend to be modeled as rational actors that behave in their own self-interests (for example, to increase income, status, or success). The models do not consider the possibility of people acting from values that are inherent to all or embracing a sense of collective responsibility for the health and integrity of the planet. For example, these models do not account for the development of social consciousness, which is defined by Marilyn Schlitz and her colleagues as "conscious awareness of being part of an interrelated community of others."[12]

Some agent-based models have suggested that beliefs can influence outcomes. For example, Jierui Xie and his colleagues showed that when 10% of a population held a committed belief (i.e., they could influence others, but their own beliefs were immune to influence), much of the rest of that population also shifted beliefs.[13] This relationship applies to shifts that can be perceived as either positive or negative from the perspective of global sustainability. How then do we calculate the probability of a thriving future for the planet, when it depends on the thousands of decisions made each day by billions of individuals? In other words, when the probability is not related to the planet, but to people taking (or not taking) action.

BELIEFS AND PROBABILITIES

Probabilities are closely linked to how we think about both predictability and uncertainty. Statistical physics is one approach that builds on the "law of large numbers" to investigate how pure randomness gives way to predictability. In *Critical Mass: How One Thing Leads to Another,* science writer Philip Ball explores how collective patterns of behavior emerge from many individuals acting independently.[14] Drawing on the science of phase transitions, which studies how matter transforms from solids to liquids to gases, Ball considers examples of people's social behavior. He notes that "a kind of regularity and order comes not from any predestination in the fates of the participants but from the very limited range of their viable choices."[15] In other words, human predictability seems to be associated with our perception of options, especially what choices we actually believe to be possible.

Uncertainty, from a quantum perspective, is not about lack of knowledge, but rather it is a characteristic of the nature of reality itself. This implies that probabilities are defined by the larger context, which includes not just the objects of interest, but human and non-human agents as well.[16] Quantum approaches to probability implicitly recognize the much wider range of possibilities and potentials for human beings to influence the social world, based on the idea that beliefs are malleable or flexible and can influence outcomes. This idea is familiar to Bayesian probability theory, which is an interpretation of probability based on reasonable expectation, which can represent either a state of knowledge or the quantification of a personal belief.[17]

To quantify the "degree" or intensity of belief, Bayesians use a formalized version of betting, whereby an agent is essentially wagering on the outcome of an event. A person may believe (or not believe) that a plant-based diet is healthier for themselves and the planet; that traveling by plane is (or is not) the only option available to them; or that fossil fuel investments are (or are not) more beneficial than investments in renewables. These "bets" may be revised based on new research, information, observations, and experiences, including conversations with others. Probabilities can then be recalculated and updated with mathematical rigor when new evidence is acquired that modifies the original degree of belief. In their research on climate change, Claudia Tebaldi and her colleagues found that using a Bayesian approach "is not only flexible but facilitates an open debate on the assumptions that generate probabilistic forecasts."[18]

QUANTUM BAYESIANISM

How does this relate to quantum social change? In *Quantum Mind and Social Science: Unifying Physical and Social Ontology*, Alexander Wendt points out that in the quantum case,

> we cannot say that there is an 'actual state of affairs' with respect to the question we want to ask, and so the probabilities refer only to possible observations. Put another way, some kind of reality is out there which gives answers to our questions, but the answers are not out there until we ask them.[19]

When a measurement or observation is made, "the probability of all the possible outcomes that are not actually observed goes to zero, and that which is observed goes to one."[20] Likewise, Karen Barad asserts that "scientific practices do not reveal what is already there; rather, what is 'disclosed' is the effect of the intra-active engagements of our participation with/in and as part of the world's differential becoming."[21] Here, she makes the distinction between *inter*actions among separate entities, and *intra*-actions among entangled elements with a larger whole, where the latter enact or materialize *perceptions* of separability. From this perspective, humans are "agencies of observation" that constantly influence the observed state of affairs and participate with nature, within nature, and *as* nature to materialize reality.[22] The future is conditional, and we are individually and collectively co-creating it through our ideas, beliefs, and actions.

As we saw in the previous chapter, Quantum Bayesianism, or QBism, is based on a "personalist" or subjective approach to Bayesian probability. As such, the collapse of the wave function is not some mysterious and unexplained event, but a result of subjective beliefs. According to this interpretation, quantum wave functions are not probabilities associated with the external world of electrons; instead, they represent subjective probabilities assigned by agents, depending on the total information available.

Emphasizing the importance of agency in QBism, physicist Hans Christian von Baeyer writes that:

> Bayesianism removes the location where probability resides from the external material world and places it instead in the mind of a person, called an agent. In this context an agent (from the Latin *agens* for "doing") is not a representative of other people but someone who is capable of making decisions and performing actions. Bayesian probability is a measure of an agent's personal *degree of belief* that an event will occur or that a proposition is true.[23]

Probabilities are thus considered numerical measures of personal degrees of belief, a dynamic referred to as "participatory realism." As described by Christopher Fuchs and Blake Stacey, "QBism is a way of investing meaning in the abstract structure of quantum theory. It tells us that quantum theory is fundamentally about agency, actions, consequences and expectations."[24] We practice, prioritize, invest in, and build things based on the beliefs that we hold about the nature of reality, and QBism makes clear the importance of integrating new information to update our beliefs and "bets" on potential future outcomes. From a quantum perspective, the future can be thought of as a "distribution of potentiality" that exists here and now.

BETTING ON THE FUTURE

Both our conscious and unconscious beliefs (i.e., those that we are and are not aware of) can influence our bets on the future. For example, many scientists enthusiastically support policies consistent with the 2°C goal of the Paris Agreement, but when asked whether they believe that this goal can be achieved, they often hesitate to answer. Some of them do not really believe that people have the capacity to change, or that it is possible to rapidly transform systems and cultures. They are not betting on a 2°C future. As another example, in some communities the belief that it is possible to limit global warming to 1.5°C may manifest as investments in solar energy installations or promotion of plant-based diets. In other communities, a belief in the inevitability of a 4°C warmer world may promote building high sea walls or introducing new forms of social protection in response to projected climate disruptions.

Beliefs about individual and collective change are constantly reinforced through various media. For example, consider a 2017 headline in *The Guardian*: "Planet has just 5% chance of reaching Paris climate goal, study says."[25] *The Guardian* headline is based on research by Adrian Raftery and colleagues, whose integrated modeling study included some important assumptions.[26] These researchers make it clear, for example, that their forecasting model was based only on past emissions. It did not explicitly incorporate future climate legislation, nor did it account for the possibility that decreasing prices for alternative energy sources could create a massive shift towards renewable energy. They argued that "this would be speculative...."[27] As with most modelling studies, theirs did not consider changes in worldviews and cultural values or an expanded sense of political agency, both of which may influence future climate policies. Individual and collective agency, especially among younger generations, plays no direct role in these scenarios.[28]

Beliefs are cognitive and "in our head," but they are much more than subjective and personal. Beliefs are also held in our bodies and reinforced by the sensations, perceptions, and emotions we experience, the judgments that we place on them, and the expectations that they create.[29] Some beliefs may be experienced as "fixed" or "solid" as a result of individual and collective traumas that affect the functioning of the nervous system and brain, inhibiting the capacity of an individual or group to thrive.[30] Becoming aware of the beliefs that we hold and reflecting on their strengths and limitations helps us to identify any blind spots, and allows us to engage more authentically with others who do not share our perceptions, interpretations, interests, and experiences. Numerous processes and practices, such as Gestalt therapy and meditation, allow practitioners to develop more flexibility and "degrees of freedom" related to their beliefs and emotions. Quantum social change calls for us to be aware of the beliefs that we hold and why, and to reflect on where we are placing our bets in every moment.

IN SHORT...

Many of the current debates about climate change are based on polarizing beliefs of either/or, right/wrong and us/them, which are often rooted in a binary, fragmented view of the world. Quantum physics challenges such "separateness" and fragmentation, and invites us to rethink the relationship between subjects and objects. As suggested by the QBist interpretation of quantum physics, beliefs influence the objective world that we perceive and co-create.[31] To create a thriving future, we may have to shift from viewing the world *through* our beliefs to looking *at* our beliefs, including how they influence the nature of our relationships, and our relationship with nature.

——— REFLECTIONS ———

- How flexible are your beliefs? Have you ever changed a belief? If so, why?

- What might a person or group have to believe to experience a limited capacity to change?

- When your beliefs are challenged by others, in which contexts do you react to defend them and in which cases do you respond by listening and engaging in dialogue?

4

RELATIONSHIPS

RELATIONSHIPS MATTER. A relationship is defined as the way that two or more things are connected, and it can be described by both the type and quality of the connections. How we relate to ourselves, each other, the environment, and the future creates patterns that tend to repeat themselves in a systematic manner, shaping both social and material structures and enabling or constraining the kinds of change we are able to achieve. Disrupting inequitable and unsustainable patterns and generating new ones that support a planet where all life can thrive is at the heart of quantum social change.

In recent decades, it has become clear that we have collectively transformed global systems in ways that threaten the well-being of all, whether by influencing land cover, species distributions, nitrogen and phosphorus cycles, or the climate system.[1] At the same time, it is often difficult for us to see how we as individuals, groups, and communities can actually influence these systems in a positive way that is aligned with both equity and sustainability. This may be an outcome of the belief that we are in a classical relationship with systems, where systems tend to be seen as separate from us, or "out there." What if we were to see ourselves not only *as parts* of these systems, in a classical, mechanistic sense, but *as systems*, from a relational, quantum perspective?

CLASSICAL RELATIONSHIPS

The systems and structures that maintain our ecosystems, economies, governments, and societies can be thought of as a collection of dynamic relationships that together form a whole. Our understanding of these dynamic relationships has led to a widespread awareness that events occurring in one location have consequences for others and for the environment, whether through the movement and spread of people, information, ideas, resources, products, or viruses.[2]

Within coupled socio-ecological systems, relationships are usually considered to be fundamentally classical, taking the form of interacting parts, entities, or "things in space." Historically, these relationships have been described with reference to clocks, machines, and more recently, computers. Within these systems, humans are generally conceived to be one among many separate and interacting parts, and their relationships and interactions are considered to be governed by classical laws of physics. In other words, social life implicitly follows Newton's laws of motion and thermodynamics, as well as the laws of electrodynamics. Humans are thus predictable, and uncertainty is related to our lack of knowledge, rather than to the nature of reality.[3]

Classical systems are inherently local, not in terms of scale, but causation; i.e., causality is attributed to something having an effect on something else. As mentioned earlier, locality implies that action at one point can influence another point *only* if something traveling through space mediates the action or influence, whether through physical processes or through energy flows, material flows, or information flows. In *Critical Mass*, Philip Ball explores how everything from traffic planning, to market analysis, to game theory and criminology can be described by mechanistic, quantitative models. In fact, complexity of form and organization can arise from simple underlying principles, "if they are followed simultaneously by a great many individuals."[4] This transition from "simplicity to complexity" brings to mind the phenomenon of fractals, which are self-similar patterns that repeat themselves at all scales, and will be discussed in a later chapter.

"Classical" social science, sometimes referred to as positivist social science, tends to model people as rational actors whose interactions with others and with the environment are genetic or learned behaviors that optimize outcomes, whether surviving, "winning," or thriving. From a classical perspective, humans have both a relationship *to* nature (e.g., they may control, manage, exploit, steward, conserve, or ignore it) and a view *of* nature (e.g., they may see nature as fragile, capricious, benign, or tolerant).[5] Humans are not, however, considered to be entangled *with* nature, or *to be* nature. Never mind that an adult human body consists of up to 60% water, that we inhale and exhale about 11,000 liters of air each day, and that our bodies contain trillions of microbial cells. Classical social science still views nature as something separate from us.[6]

In today's world, most systems have been designed or developed to be resilient and adaptive to some degree of change. This applies to energy systems, transport systems, agricultural systems, water systems, health care systems, insurance systems, and so on. Many people, institutions, and groups benefit directly or indirectly from the stability and predictability of these systems; consequently, there are numerous controls in place to maintain business-as-usual trajectories, while incrementally adapting systems to observed and projected changes in the population or environment. Beneficiaries of these systems risk great losses if the systems break down or are deliberately disrupted or transformed. It is not surprising that the more power and influence a beneficiary derives from these systems, the greater their interest in preserving the status quo.

Social and environmental activists and change agents are often complicit in maintaining the very systems they want to change, sometimes without realizing it. Drawing on insights from psychology and marketing, mathematical models have been developed not only to predict, but also to regulate, manipulate, optimize, and control behaviors and outcomes. In fact, understanding and influencing collective patterns of behavior has become a lucrative industry in the era of algorithms, artificial intelligence, and "big data." This manipulation is driven by particular understandings of human perceptions, behaviors, and decision-making. However, as Cathy O'Neil writes in *Weapons of Math Destruction: How Big Data Increases Inequality and Threatens Democracy*, "mathematical models, by their nature, are based on the past, and on the assumption that patterns will repeat."[7] To generate transformations to an equitable and thriving world, we will have to disrupt some of these repetitive patterns and intentionally establish new ones.

QUANTUM RELATIONSHIPS

How would we relate to systems that are fundamentally quantum, i.e., characterized by entanglement, uncertainty, indeterminism, and potentiality? In quantum systems, there is nothing that is "apart" or fully separate, and if we consider the universe to be one large quantum system, this includes humans and non-humans alike. If we are living in an entangled universe, our understanding of social change should take this into account. This includes our understanding of the relationship between social structures and human agency – an area of inquiry that has long been of interest to social scientists.

Karen Barad's concept of agential realism considers structures as "material-discursive" phenomena, which are the basic units of reality.[8] For Barad, saying something is so does not make it so; it is through intra-active practices that the material world we experience comes to be. In her view, these practices establish the boundaries, properties, and meanings of the world; structures materialize through our intra-actions within one large, entangled system. [9] What might this look like in reality? As an example, consider the concept of carbon footprints, which is a measurement of the total greenhouse gas emissions attributed to an individual, organization, product, service, or industry. This concept has generated material phenomena such as carbon accounting, carbon calculators, carbon budgets, and carbon trading. It has also influenced behaviors, with some people choosing to drive less, fly less, consume less, or eat plant-based diets to reduce their personal carbon footprints.

Language matters too, in a literal sense, with grammar and words also influencing the structures and meanings through which we interpret the world. In *Wholeness and the Implicate Order*, David Bohm describes language as "an undivided field of movement, involving sound, meaning, attention-calling, emotional and muscular reflexes, etc."[10] In fact, Alexander Wendt's "quantum social theory" considers language and linguistic meaning to be a quantum social phenomenon. His starting point is the debate over whether language is a cognitive construct in our heads or whether it is constituted and regulated by norms shared by a community of speakers.[11] Wendt points out that both arguments within this debate are implicitly linked to a classical frame of reference and subject-object dualism. A quantum frame of reference, in contrast, is holistic and emphasizes the significance of both context and entanglement.

Meanings are shared or "entangled" as words are activated through speech acts that involve not only the speaker and the meaning that they intend to communicate, but also the listener and *their* experience and association with words. Wendt points out that in communication, a shared understanding will depend on how the words interact with the listener's memory of words and their associations, which may be very different from the speaker's.[12] For example, although the concept of a carbon footprint is mutually understood by people concerned about global warming, it may be incomprehensible to those who are skeptical of climate change, or unfamiliar with the jargon of climate change mitigation. Since multiple potential meanings "collapse" into an actual meaning through speech acts, a shared context is critical to successful communication.

Context both defines and influences the nature and quality of relationships that make up a particular system. In quantum systems, superpositions refer to indeterminate states that represent a wide spectrum of potentialities. Alexander Wendt considers social structures as *"superpositions of shared mental states – social wave functions."*[13] This contrasts with the classical view, whereby people hold a portfolio of mental states in their head, which are then the basis for their actions. In Wendt's view, "social structures are continuously popping in and out of existence with the practices through which they are instantiated."[14] In other words, social norms, rules, regulations, and institutions are not "really real" in the sense of having a physical basis, but constructions that materialize through language, shared meaning, identities, beliefs, and values.

This points to the role each and every one of us plays in propagating the patterns and relationships that are degrading the environment, diminishing people, eradicating species, oppressing labor, perpetuating warfare, and so on. Yet it also points to the profound role we can each play in generating structures that promote sustainability, equity, and peace. As Wendt argues "There *is* no higher level in social life above that of individuals: the reality of social life is flat."[15] One can therefore think of structures as "being *pulled* out of the quantum world of potentiality into the classical world of actuality by agents."[16] This is not an appeal to classical individualism, which assumes fully separate agents, but to a holistic and relational individualism characterized by non-local entanglements that are mediated by language.[17] Karen Barad's agential realism similarly suggests that the connections among humans and other species, objects, and events extend beyond the metaphysics of individualism, in that "Events and things do not occupy particular positions in space and time; rather, space, time, and matter are iteratively produced and performed."[18] According to Barad, this production and performance of events and things is linked to intra-actions, or the dynamics of material practices.

The quantum nature of our social structures can be seen in the development of social and cultural norms and their material consequences. For example, similar to the traditional Roman Catholic practice of not eating meat on Fridays as a sign of universal penance, the concept of "meatless Mondays" for the sake of the environment creates options and alternatives that would not otherwise be available, and which may eventually contribute to new social norms, institutions, and menu alternatives that support plant-based diets. The structures created by society, including many cultural norms and institutions related to governance, are thus often not real, but artifacts of traditions, shared beliefs, values, expectations, and intentions. Becoming aware of alternative possibilities and how they come to matter is an important aspect of quantum social change.

THE LANGUAGE OF RELATIONSHIPS

We currently lack nuanced words – especially in languages such as English – to describe relationships associated with quantum systems, such as nonlocality, complementarity, and entanglement. Although a quantum system is holistic, it is not made up of wholes and parts, at least not in the classical sense that we associate with a mechanistic worldview. This is difficult to grasp, for even though the world is shaped by quantum phenomena, we still tend to describe and discuss them from a classical perspective. This is not surprising, as classical physics has formed the basis of modern culture and is part of our everyday experiences. In *Quantum Language and the Migration of Scientific Concepts*, Jennifer Burwell discusses how difficult it is "to invent a new mode of narration appropriate to quantum physics when it is not just a matter of finding the right terms, but rather of discovering a new grammar."[19]

To overcome the dualism inherent in the classical worldview, we need both a new grammar and a new vocabulary. Fortunately, alternatives are available and being developed. For example, philosopher Ken Wilber suggests that reality can be conceived as being composed of not things or processes, but of *holons* or "wholes that are simultaneously parts of other wholes, with no upward or downward limit."[20] Wilber describes the interiority of individual holons to be something that is "in essence the same as *consciousness*."[21] He further elaborates that:

> since reality is not composed of wholes, and since it has no parts – since there are only whole/parts - then this approach undercuts the traditional argument between atomism (all things are fundamentally isolated and individual wholes that interact only by chance) and wholism (all things are merely strands or parts of the larger web or whole) ... There are no wholes, and there are no parts. There are only whole/parts.[22]

Linguistically, the *duality* of whole/parts captures the relational interdependence, complementarity, and lack of separation between the two concepts, distinguishing it from the *dualism*, i.e., wholes and parts as two distinct and opposing concepts.[23] This is not to be confused with the philosophical and spiritual understanding of the concept of non-duality as an experience of oneness. Since the English language currently does not have the words to express the nuances of quantum relationships, a forward slash and square brackets will be used here as an interim symbol to communicate relational interdependence, such as [whole/parts], [I/we], and [both/and]. While this may appear inconsequential, symbols can create deep roots in the unconscious and diffuse an emotional quality into situations and behaviors, or in Edward Sapir's words: "[The symbol] expresses a condensation of energy, its actual significance being out of all proportion to the apparent triviality of meaning suggested by its mere form."[24] The brackets are a simple way of communicating the essence of entangled relationships, helping us to shift from the "us versus them" perspective that perpetuates a fragmented and divided society, towards an [I/we] perspective that acknowledges our inherent oneness.

This perspective is increasingly recognized in academic research, and has a long history in many cultures and wisdom traditions. For example, clinical psychiatrist Daniel Siegel uses the term MWe ("Me" plus "We") to describe this intra-connected relationship.[25] In *Sand Talk: How Indigenous Thinking Can Save the World*, Tyson Yunkaporta uses a pronoun that translates in English to "us-two," and he explains that in Aboriginal society, "languages are expressions of land-based networks and facilitate communication across all of these individual nodes and collectives of nodes within and between systems."[26] This perspective is also expressed in the concept of "right relations," a term used in many indigenous communities to refer to what human geographer Irmelin Gram-Hanssen and colleagues describe as "an obligation to live up to the responsibilities involved when taking part in a relationship – be it to other humans, other species, the land or the climate."[27]

Right relations reflect a quality of relationship that is aligned with Karen Barad's view that "subjectivity is not a matter of individuality, but a relation of responsibility to the other."[28] How we perceive and engage in these relationships matters, quite literally. Fortunately, as Yunkaporta points out, we can look to other cultures for inspiration on grammar and vocabulary to describe entangled systems and relationships. For example, in *An Ecotopian Lexicon*, John Esposito offers *in lak'ech—a la k'in*, a Mayan exchange where one person greets another with "I'm another you" and the other person responds with "You're another me."[29] According to Esposito, this exchange "affirms an irrefutable bond between the interlocutors that is at once psychological and spiritual" and highlights the primacy of relationships based on solidarity and coexistence. [30] Such words help us to capture the [I/we] and [whole/parts] dimensions of mattering.

IN SHORT...

Our beliefs about the nature of reality influence our relationships, including our relationship to nature. For many of us, our relationships to concepts and ideas such as *self*, *other*, and *nature* have been influenced by a classical, mechanistic view of the world. Relationships within a quantum system are characterized by entanglement, uncertainty, indeterminism, and potentiality. In quantum systems, there is nothing that is "apart" or fully separate, and if we consider the universe to be one large quantum system, this includes humans and non-humans alike. When we consider Karen Barad's concept of intra-actions, together with Alexander Wendt's perspective on language and linguistics as quantum social phenomena, it becomes clear that who we are, what we do, and how we relate and communicate matters. In short, [I/we] matter and the words and metaphors we choose and use have the power to transform both our relationships and our shared reality.

—— REFLECTIONS ——

- What relationships are most important to you in your everyday life?

- Do you feel you have any power to influence "systems" (e.g., health, food, waste, education, or economic systems)? If so, how?

- The last time you made a significant change in your life, how did it affect your relationships (e.g., to family, friends, work, time, the future, etc.)?

5

METAPHORS

METAPHORS MATTER. Regardless of whether or not we think that quantum physics is relevant to the social world, it does matter, metaphorically. Metaphors represent a figure of speech, whereby one thing is used to mean another, as in the use of the term "greenhouse effect" to explain how our planet is warming. Metaphors are powerful, in that they create meaning and convey images, emotions, values, and judgments about what is true or false, real or unreal, and possible or impossible. In fact, in *The Metaphors We Live By*, George Lakoff and Mark Johnson argue that "New metaphors have the power to create a new reality."[1] As they explain, this occurs when we start to make sense of our everyday experience in terms of a new metaphor, and especially when we begin to act in accordance with the metaphor. In this way, it eventually becomes part of our experienced world.[2]

Issues such as climate change are already discussed largely through metaphors, yet most of these are still based on a classical and fragmented representation of the world. In an analysis of metaphors in 63 international science-policy reports related to climate change mitigation, Christopher Shaw and Brigitte Nerlich found that there is a tendency to universalize climate change impacts into an impacted/non-impacted dichotomy, reducing climate change to a problem that can be solved through economic cost-benefit analyses.[3] Such metaphors constrain the discourse and limit the solution space by marginalizing alternative policies. They shape the way that we respond, or do not respond, to climate change.

COMMUNICATING WITH METAPHORS

Metaphors can be powerful in communicating facts, ideas, under-standings, and meanings from one person or group to another, and from one to many. They are considered an indispensable tool for practicing and communicating science. In *Science, Order, and Creativity*, David Bohm and F. David Peat consider metaphoric perception to be fundamental to science, in that it brings together what were once believed to be incompatible ideas in radically new ways.[4] They describe the metaphor as an act of creative percep-tion that occurs when a state of intense passion and high energy dissolves the rigidly held assumptions associated with commonly accepted knowledge. However, metaphors can also be used strate-gically to influence policy. For example, Cynthia Taylor and Bryan Dewsbury point to experimental studies that show how the framing of issues such as climate change through metaphors "can subtly, and covertly, influence the perception of risk, the sense of urgency, and the level of support for proposed 'solutions' by acting on pre-existing cognitive schemas and prompting affective responses."[5]

Climate change metaphors such as "tipping points," "hothouse Earth," or "a 4°C warmer world" are intended to convey urgency and inspire action, but could they be inadvertently contributing to hopelessness, despair, and inaction? It is not a question of the degree of truth represented in the metaphor, as Lakoff and Johnson emphasize, but rather how it is perceived and acted upon: "We draw inferences, set goals, make commitments, and execute plans, all on the basis of how we in part structure our experi-ence, consciously and unconsciously, by means of metaphor."[6] For example, frequent use of the metaphor "time is money" leads us to talk about spending time, saving time, using time, borrowing time, running out of time, and wasting time. Time has become a valuable commodity, and money is the metric that matters. Given the prev-alence of this metaphor, it is not surprising that many feel that we have neither the time nor the money to address climate change.

Metaphors are embodied and deeply inscribed into our language, and it is not easy to simply replace them. Bohm and Peat refer to our common tendency to unconsciously defend ideas that are considered significant to "the mind's habitual state of comfortable equilibrium."[7] They point out that it takes more energy to change one's mind than to maintain the same pattern of thought. Rather than opening our minds to new ways of seeing and being in the world, we tend to retrofit new ideas into existing frameworks and describe them with familiar language. This may be the usual approach, but we live in a time that compels us to think differently. Fortunately, as Lakoff and Johnson emphasize, we do have considerable cognitive flexibility and a limited yet crucial freedom when it comes to conceptualizations.[8] Furthermore, they stress that we have multiple metaphors for our most important concepts, and that the significance of metaphors can sometimes be reprioritized.

Challenges to established thinking often occur when a situation becomes unbearable. A convergence of crises may encourage people to deliberately adopt metaphors that contribute to new patterns of thought, opening up the potential for more equitable and inclusive ways of responding to global change. Perhaps the more frequent use of quantum metaphors can help us both to relate to an abstract problem like climate change in a different way, and also to respond to it differently. For example, would we relate differently to ourselves, to others, to the environment, and to the future if we thought of ourselves as "nonlocally entangled" and full of potential? How would we respond if we believed that we really mattered?

QUANTUM METAPHORS

One of the most common quantum metaphors in use today is the *quantum leap*. This describes an abrupt and large change, as in "we need to make a quantum leap in renewable energy production to address climate change." This is an imperfect and misleading metaphor, as the mere use of the word "leap" suggests a physical jump through space and time.[9] In quantum physics, the term actually refers to a seemingly discontinuous change of the state of an electron from one energy level to another in an atom or molecule. To talk about making a quantum leap to sustainability moves us into unknown territory, for it does not appear to be a linear trajectory or pathway, but rather a shift in energy and vibration. Although we can look for the early signals of tipping points to indicate phase changes in the classical physical and social world, in the quantum world we do not yet seem to know exactly where and when these leaps will happen. As Karen Barad points out, a true quantum leap to sustainability is an "intra-play of continuity and discontinuity, determinacy and indeterminacy, possibility and impossibility that constitutes the differential spacetimematterings of the world."[10] Barad is suggesting that quantum leaps occur in liminal or in-between spaces that have no direct correlates in our classical conceptualization of reality. Perhaps this "space between things" transcends our individuality and is about connection. Whether we refer to this liminal space as [I/we] or [whole/parts], it can be a powerful source of quantum social change.

The quantum concept of *superposition*, though unfamiliar to many people, can serve as a useful metaphor for indeterminacy and potentiality. As mentioned earlier, quantum superposition refers to all possible states of a phenomenon existing at the same time, in the form of a probability wave that expresses potentiality. According to many interpretations of quantum physics, a quantum wave function is not real – it is only an abstraction for all possibilities existing at once, each with the potential to exist as an actual state when a measurement is made and the wave of potential collapses into a defined state. The probability of any outcome is undetermined yet not random; unlike the deterministic equations of classical physics, the Schrödinger Equation describes the evolution of the quantum wave in space and time by squaring its amplitude for each point. Karen Barad makes it clear that "Superpositions challenge our classical metaphysical view of the world," representing instead the indeterminacy of the quantum world.[11]

What does "being in superposition," or "taking a superposition," mean for social change? First and foremost, it draws attention to possibilities and alternatives, and to the potentiality that exists in every moment. Recall that Alexander Wendt considers social structures to be social wave functions, or "superpositions of shared mental states."[12] This means that the norms, rules, institutions, and ideas that we have created to organize and manage systems, such as those related to water access, air quality, health care, food, or social security, represent one among many possibilities. In discussing the implications of superpositions, Wendt makes two important points. Firstly, similar to quantum wave functions, social structures are not "real" in the classical sense. Rather, we experience them as real when potentialities collapse into a particular outcome or condition. Secondly, social structures do not exist above and beyond individuals and their practices; they emerge from the entanglement of mental states among those whose ideas and practices create them.[13] In other words, social structures are indeterminate until a speech act or some other type of action is taken. For example, public-private partnerships governing water access in times of drought are not "real," rather they are practices created by shared mental states that have material manifestations and consequences. Quantum superposition draws attention to the many equitable and sustainable alternatives that exist and can be "collapsed" into reality, right here and now.

There are probably many other quantum metaphors that can be used to describe a holistic, entangled system that is full of potential, such as *quantum coherence* or *nonlocality*. Or perhaps "many worlders" and "hidden variables" will provide new ways of talking about peculiar social phenomena. There is, of course, always a danger of misusing metaphors, or of using them superficially rather than substantively.[14] For example, consider entanglement, which is widely used in the classical sense to mean twisted together, intertwined, or caught up in something. As we will see in the next chapter, the reality of quantum entanglement represents far more than a metaphor. It directly challenges classical ideas of causality and has implications for both meaning and mattering.

RESONANCE

Metaphors matter because they invoke images and meanings that can help us to make connections that otherwise would elude us. They can create a resonance among people and "carry" new ideas into the world. However, to come up with new metaphors requires creative thinking, and this is where analogies come in. Analogies are based on logical arguments that draw attention to similarities between different things. The use of analogies can be an effective way to trigger insights into physical and social relationships.[15] They build on existing metaphors and provide us not only with ways of seeing similarities and differences, but also with a means for free and improvisational thinking. For example, analogies between physics and jazz music may provide useful insights into quantum social change.

Stephon Alexander draws such an analogy in *The Jazz of Physics*, where he defines resonance as "the means through which vibrational energy can get transferred from one physical entity to another with great efficiency,"[16] He emphasizes the power of this when it comes to the transfer of energy: "It is through vibration, resonance, and interaction that the microworld is linked to the macroworld."[17] Of course, all analogies introduce limitations, but Alexander points out that this also introduces an opportunity to discover something new.[18]

As an example, in Western philosophy, there is a tendency to think of individuals as discrete units, somewhat analogous to individual marbles, with society analogous to a box of marbles, where individuals of different sizes, colors, or patterns come together in groups, together yet still separate. Quantum social theory reveals the limitations of this particular analogy, maintaining instead that individuals are analogous to *both* particles *and* waves. As with particles, the characteristics of an individual (e.g., height, weight, skin, hair, and eye color, etc.) can be observed and measured, but the meaning or significance of these characteristics is always relative to the norms that are deemed desirable or acceptable by the collective, i.e., the social wave function. A social wave function that recognizes the value of diversity will collapse into a very different reality than one that emphasizes homogeneity. Similarly, a social wave function that values equity and sustainability will manifest a different world from one that emphasizes greed and exploitation.

As social norms change, so do the meanings of words and their significance. What was once an acceptable expression in one place or time may be considered unacceptable at another. It resonates differently with the social wave function, creating an uncomfortable dissonance for many. Dissonance and resistance are to be expected as social norms shift; this is part of transformative change processes. Returning to the analogy between physics and jazz music, physicist Stephon Alexander describes the musical character of a universe teeming with structure that is sustained through a "dance among harmony, symmetry, instability, and the gaps of improvisation."[19] With reference to the uncertainty principle, he goes on to reflect on the profound implications of the fact that we can never "pin down" both the particle-like and wave-like properties of a quantum entity. Interestingly, he likens this to musical improvisation, suggesting that the more *certain* a saxophone player is about playing the note she/he is about to play, the more possibilities open up for the notes that follow: "Every improvisation is a new experience – not a reiteration of something past but of something never done before."[20] Put another way, when we act in the moment, with an awareness and intention that is based on the recognition that [I/we] are connected, we create space for new possibilities to emerge.

In short...

Metaphors are powerful figures of speech, and they have the potential to both shape and communicate a new reality. They influence what we perceive and experience, and also how we act. Metaphors such as quantum leaps and superposition offer an opportunity to express processes and relationships in new ways, creating a resonance that is analogous to a vibrational energy. The energy embedded in quantum metaphors may be critical to improvising social change in an entangled world. Particularly when we recognize that quantum entanglement is not just a metaphor, but the nature of reality.

——— Reflections ———

- What metaphors do you use on a day-to-day basis to describe climate change?

- Have you ever consciously adopted a new metaphor to describe a situation or possibility? (For example, "hothouse Earth" has recently been introduced to describe the risks associated with climate change feedbacks).

- "If we were tiny, we would see our inner waviness," writes physicist Stephon Alexander in *The Jazz of Physics*.[21] What type of music would you imagine that your inner waviness is playing right now?

6

ENTANGLEMENT

ENTANGLEMENT MATTERS. Moving beyond metaphors, quantum entanglement has a specific meaning in quantum systems, where information about separated particles is revealed when a measurement is made on one of them, even though no information passes between them. This non-local relationship can be described as correlated (or "co-related") rather than causal. Entanglement is a real phenomenon that provides a fruitful starting point for exploring how we both relate to and engage with social change.

This quantum version of "connected and correlated" defies our classical understanding of relationships. Entanglements, as Karen Barad writes, *"are uniquely quantum mechanical—they specify a feature of particle behavior for which there is no classical physics equivalent."*[1] This challenges many of the assumptions we hold about the nature of reality. Insights from quantum social science point to the significance of entanglement for understanding relationships between individual change, collective change, and systems change. If we take this one step further and consider that our entanglement is not only with each other, but with all species and the environment, then boundaries begin to blur. As Barad puts is, "Boundaries, properties, and meanings are differentially enacted through the intra-activity of mattering ... The very nature of materiality is an entanglement."[2]

CLASSICAL ENTANGLEMENT

Entanglement, derived from the verb "entangle," can be traced back to the Middle English (11th to 16th centuries A.D.) word *entanglen*, which means to involve in difficulty or to embarrass. In common usage today, entanglement describes a state of being twisted together or entwined. It can also describe a complicated or compromising relationship or situation, including climate change. Such understandings of entanglement imply "connected yet separate." Indeed, in modern society, relationships between individuals and groups, humans and the environment, and nature and society are premised on the idea that there is a clear separation between entities.

Within the modern Enlightenment worldview, entanglement conjures an image of normally separate elements now being enmeshed, and this sometimes has negative connotations. We may, for example, be entangled in an affair, in politics, or in a bad business deal. Or entangled with unhealthy thought patterns or behaviors. Entangled relationships among separate entities are often based on interests, identities, emotions, values, or force. Still, not all entanglements are negative; we can be entangled in neutral or positive ways as well. For example, an empirical study carried out by Ivana Konvalinka and colleagues found that the heart rates of close friends and relatives of participants in a Spanish fire-walking ritual exactly matched (or synchronized with) those of the firewalkers, but that there was no correlation between the firewalkers and other spectators who did not know them.[3] This synchronous arousal was not mediated by a direct exchange of matter or energy, thus the researchers conclude that there seems to be some informational basis to the coupling.

Some examples of similar "self-other" connections have been shown to result from mirror neurons in the brain, generating a response known as affective empathy.[4] Research on the social nature of the mind shows that social relations and connections are fundamental to human well-being. Matthew Lieberman and Naomi Eisenberger found that social and physical pain activate the same regions of the brain, which draws attention to "the critical importance of the social world for the surviving and thriving of humans."[5] This relational view contrasts with classical understandings of human nature. As Lieberman notes:

> The theorizing of Hobbes, Hume, and other intellectuals who claim that self-interest is the source of all human motivation has produced a self-fulfilling prophecy. Their theory and everyone who repeats it has affected how the rest of society behaves. Because we have been taught that people are self-interested, we conform to this cultural norm to avoid standing out.[6]

As discussed earlier, the beliefs we hold, repeat, and reinforce influence how we collectively relate to and organize society.

We are also entangled with the environment, as exemplified in the concept of *coupled* social-ecological systems. This is perhaps most clear in our relationship with the climate system: the air we breathe includes both oxygen and carbon dioxide; plants take in carbon dioxide and release oxygen, storing the carbon as biomass. Burning vegetation and fossil fuels releases stored carbon into the atmosphere. In the atmosphere, these molecules of carbon dioxide absorb infrared radiation, vibrate, and produce heat, resulting in warmer temperatures. This creates a troublesome situation, since a warmer atmosphere influences all types of processes, cycles, flows, and states, including winds, currents, oceans, ice, soils, vegetation, and ecosystems, not to mention the physiology of humans and nonhuman species. We are entangled in a process where we are transforming the very systems that we depend upon to maintain a thriving planet.

This image of classical entanglement is well fitted to the economic and political relationships prioritized within dominant modern worldviews, where we often find ourselves reduced to individual consumers, with a focus on "I" more than "we" and with little sense of the extent of our connections and our potential for collective agency. For many, life seems to be primarily determined by circumstances and systems that we cannot influence. The global economy, for example, has been organized according to neoclassical economic theory, prioritizing individualism, capitalism, and profit over the many alternative ways of relating to and managing ourselves, society, the economy, and nature. Below, we will consider a quantum perspective on entanglement and what this means for our understanding of social change.

QUANTUM ENTANGLEMENT

Quantum entanglement is based on the idea of nonlocality. As mentioned earlier, nonlocality describes correlations that are *not* associated with a chain of causal mechanisms propagating continuously through space, from one point to the next. Albert Einstein was skeptical of entanglement, since it implied that information can travel faster than the speed of light, and he referred to it as "spooky action at a distance." However, in physicist Nicolas Gisin's view, entanglement is not about action at all: "there is no cause here that acts over there. Rather, entanglement is a kind of 'probabilistic cause' whose effects can show up in several places, without allowing communication at a distance."[7] Confusing entanglement with causality does not make sense from a quantum perspective, a point often overlooked in interpretations suggesting that consciousness causes the collapse of the wave function.

Quantum nonlocality is a robust concept that seems difficult to grasp from our current frame of reference. However, is it really so novel? Gisin points out that Newton's theory of universal gravitation also involved nonlocality, in that shifting a stone on the moon would immediately influence the weight of an object on planet Earth. Such nonlocality is experienced every day when the moon's gravitational pull influences the tides of the oceans and other water bodies. The nonlocality of gravity was treated as an ultimate truth for centuries, until 1915, when Albert Einstein developed his general theory of relativity. It is therefore peculiar that, in Gisin's words, "We had Newton's nonlocality up until 1915, and we have had quantum nonlocality since 1927. So apart from a narrow window of just twelve years, physics has always been nonlocal."[8]

The idea of quantum entanglement was initially only a thought experiment, but in recent decades it has been verified experimentally again and again, over increasing distances and across multiple systems.[9] For Karen Barad, entanglement implies that "The very nature and possibilities for change are reworked."[10] Quantum physics has thus challenged traditional notions of causality and conceptions of dynamics linked to external forces. To understand how entanglement influences social change, we will now look more closely at the idea of language as a quantum phenomenon.

ENTANGLEMENT THROUGH LANGUAGE

Language has always had profound implications for how humans relate to each other and to the environment. The language of reductionism, individualism, and determinism has penetrated our social world and limited our ability to perceive nonlocal connections and entanglement. David Bohm notes that it can be difficult to see alternative worldviews within the existing structure of our language and meaning-making.[11] For example, slaves were previously legally regarded as "chattel" or the disposable property of their "owners" – a language that legitimized a worldview where it was permissible to violently abuse people and deny them legal status and rights.[12] As another example, many indigenous groups did not have any concept or language of ownership of the land they lived on. This was in contrast to empire-building nations that did have such a language, as well as legal concepts to support it. Some scholars claim that this legitimized the argument that such lands did not have any "owner," and thus could be claimed. In their (world) view, they were "settling" these lands, rather than conquering and expropriating them.[13]

Quantum social science tells us that individuals are always "indeterminate" and full of potential. Yet as mentioned earlier, we seem to lack the means to express this potentiality clearly and convincingly, at least through modern languages and the worldviews that they embed and perpetuate. Philip Ball reinforces this idea in *Beyond Weird: Why Everything You thought You Knew about Quantum Physics is Different*, pointing out that "There's no unique way to express what entanglement is. We lack words to convey such concepts with precision and clarity, and so we need several different ways of looking at them before we can begin to grasp what they are about."[14] Approaching entanglement through language allows us to consider that "meaningful" conversations can influence *how* we engage with social change: when we are aware that we are entangled within a quantum system and part of the phenomena that we are measuring, we have greater potential to open up inquiries with others and generate insights that are truly transformative.

According to Alexander Wendt's quantum social theory, we are non-locally correlated with each other through language, speech acts, and meaning. [15] Concepts, as Wendt describes them, typically have many meanings and lack definite properties in the abstract. He argues that potential meanings are always superimposed, with each represented by a different vector in the wave function, each with different weights or probabilities. For example, "collaboration" can mean working with someone to produce something, or it can mean traitorous cooperation with an enemy.[16] This means that we can talk about collaborating to develop alternatives to fossil fuels, or we can talk about collaborating with oil companies through continued financial investments. Although concepts can have multiple meanings, the meaning becomes specific when expressed in a particular context.

We are embedded in a quantum system, and "by virtue of our entanglement from birth in social structures, human minds are not fully separable."[17] In describing language as a quantum phenomenon that is entangled through meaning and context, Wendt considers that concepts are not "objects" as described by an orthodox view, but "processes that unfold in time."[18] As such, we can consider ourselves as "walking wave functions." What does this mean? It means that we are not fixed and determined, but full of possibility and potential, and nonlocally connected with others through language and shared meanings. From the perspective of quantum social change, an awareness of how language connects – and how it resonates across space and time – moves us beyond waiting for a "critical mass" to manifest change. Rather, it shows that each of us literally "matters."

Q METHODOLOGY

This association between language and quantum entanglement is not a new one. In fact, it is the underlying basis for Q Methodology, a mixed method used in research for assessing subjective attitudes towards a subject or claim. It is based on a sorting of statements selected from a sample drawn from the larger social context, or "cultural field."[19] Q Methodology was developed by psychologist and physicist William Stephenson in the 1930s, a time when the field of quantum physics was being intensely developed and discussed. Stephenson recognized that the mathematics of the statistical factor analysis used in psychology was the equivalent of the matrix model of quantum mechanics.[20]

In Q Methodology, language is recognized to be quantum coherent (i.e., a probability wave function), collapsing into meaning when it is expressed. Q Methodology is based on a collection of statements about a topic, gathered from the full range of views on the topics. The statements are then "sorted" by participants according to a Likert scale, ranging from "most disagree" to "most agree." A statistical factor analysis is then performed in order to identify significant views or discourses. In a Q study, the statements are considered passive and meaningless until the sorting process expresses an individual's feelings, conceptions, and concerns about them. This means that it is *the communicative contribution provided by a person in Q sort form* that acts as the quantum phenomenon."[21]

An individual's view is closely tied to that of a social group, and the power of social and cultural norms cannot be understated. As Simon Watts and Paul Stenner write, "processes of self-reference are indelibly tied to *simultaneous processes* of cultural reference."[22] Given the power of culture to define a discourse, they point out that "change at this macroscopic level demands a wholesale shift in consensus amongst the population at large. Cultural change requires that large numbers of individuals refuse the conventional *perches*…"[23] Such perches represent the socially-acceptable views and perspectives that we are conditioned to believe, and the pressure to conform to acceptable representations and viewpoints of the majority is considerable. Since social connectivity is essential to well-being, an individual's freedom to make choices is often constrained by what is deemed culturally acceptable.[24] Watts and Stenner emphasize that pre-existing narratives and the conceptual structures of the cultural field shape "our understandings, imaginings, the stories we tell, the accounts we find plausible, the nature of our interactions, and so on."[25]

Quantum social change requires that we step outside of our pre-existing narratives and instead have the courage to shift social and cultural norms to support an equitable and thriving world. Fortunately, it is possible to intentionally use language and meaning to connect with the wider cultural field and create a new shared context, and this is most often done by telling new stories about ourselves and our future.

Our Entangled Future

Stories play an important role in transforming social, political, economic, ecological, and intellectual worlds, both through their descriptive and normative elements, and through the metaphors and meanings that are communicated. They have the power to challenge our thinking, engage us with new perspectives, and disrupt the lines that we have drawn between "us" and "other." As geographer Emilie Cameron writes, storytelling "orients itself toward the emergent, the not-yet-here, and participates in the materialization of new realities."[26]

Classical stories in both the arts and the sciences portray understandings of human-environment relationships through diverse plots. Jonathan Phillips identifies eight of the most common plots told by Earth System scientists: cause-effect, genesis, emergence, destruction, convergence, divergence, oscillation, and metamorphosis.[27] He identifies the "cause-effect" plot to be the most common storyline, and considers it a fundamental motif in geosciences. Unfortunately, when a tragic apocalyptic rhetoric emphasizing a catastrophic end-point based on a linear, cause-effect plot dominates media coverage of climate change, the potential for altruistic human agency is diminished.[28] Indeed, humans are often the villains in stories of global environmental change, and seldom the heroes.

To address climate change, biodiversity loss, global inequality, and other sustainability issues, we have to tell different and more meaningful stories. Cultural historian Thomas Berry emphasizes this point:

> It's all a question of story. We are in trouble just now because we do not have a good story. We are in between stories. The Old Story - the account of how the world came to be and how we fit into it - is not functioning properly, and we have not learned the New Story.[29]

In *Climate: A New Story*, Charles Eisenstein points out that "The problem with the mechanistic view of nature-as-thing is not only that it numbs our compassion and facilitates our plunder. It also cripples our ability to serve as agents of positive transformation."[30] Fortunately, new stories are being imagined and told, often invoking powerful metaphors within plots that emphasize agency and connection.

As an example, a call for contributions to *Our Entangled Future: Stories to Empower Quantum Social Change* produced nine short stories that describe entanglement, human and non-human agency, and the potential for small changes to make a difference.[31] The stories are neither utopian nor dystopian, but instead they highlight both the wisdom and potential that exists, right here and now, for generating transformative change. For instance, Chris Riedy, author of *The Witnesses*, was inspired by the quantum concept of entanglement and the idea that if we could become more aware of the way that our lives were entangled with others, our collective values would start to shift. His protagonist's sense of "not mattering" at the beginning of the story was clear, and it is something many people can relate to: "We knew something had to change but we had no idea how our actions could make any difference."[32]

The stories we tell influence the reality that we perceive and experience, often unconsciously. Yet stories do not exist only in our minds; we also embody them. In *The Body in the Mind: The Bodily Basis of Meaning, Imagination, and Reason,* Mark Johnson emphasizes that the mind and the body have never been separate, and that meaning and rationality are grounded in recurring structures of *embodied* human understanding: "*Understanding is an event* – it is not merely a body of beliefs (though it includes our beliefs). It is the means by which we have a *shared, relatively intelligible world.*"[33] This influences how we perceive and engage with the world, or as political scientist Laura Zanotti writes, "the ways we imagine how the world is made shapes what we consider politically possible."[34] Consciously telling a different story may be critical in enhancing individual and collective agency to generate transformative change. However, to "consciously" tell such a story, one has to believe that consciousness exists, that it can change, and most of all, that it matters.

In Short...

Quantum entanglement describes a relationship that is non-local and correlated (or "co-related"), rather than causal. This is quite different from our classical notions of connectedness, and has profound implications for the ways that we think about change. Though quantum entanglement is not a visible characteristic in our "noisy" everyday world, we can experience it through language, as proposed by Wendt's quantum social theory and operationalized in Q Methodology. This is based on the idea that while concepts typically have many meanings and lack definite properties in the abstract, specific meanings emerge when we communicate. Being conscious of the language that we use, how we both speak and listen, and the stories that we tell enables us to relate to change in a different way, which opens up new possibilities for conscious social change.

—— Reflections ——

- Have you ever felt or experienced non-local entanglement, i.e., a connection with another person, species, place, object, or phenomena that transcends space and time?

- Are there any types of culturally-accepted norms and beliefs that you have deliberately challenged in the name of sustainability? Was there any pushback from your family, friends, or colleagues?

- If you were to write a short story about "Our Entangled Future," what would be the most important quality of your main character?

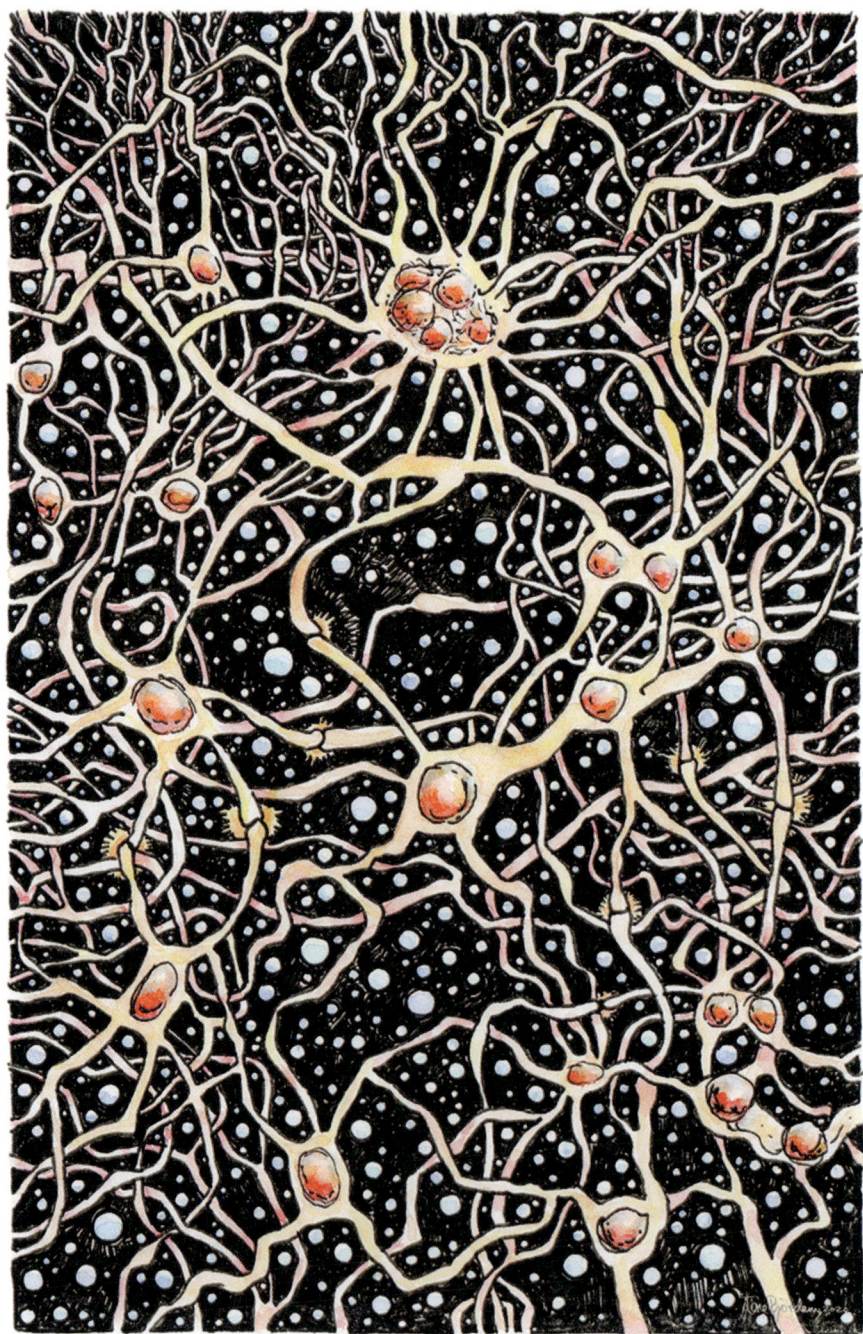

7

CONSCIOUSNESS

CONSCIOUSNESS MATTERS. In fact, what we believe about consciousness has a lot to say about how we relate to social change. What exactly is consciousness? Does it exist? Can it develop over lifetimes and generations? Consciousness has been defined as the experience, awareness, or perception of something. It has also been described as a self-awareness or reflexivity regarding one's relationships to self, to other, to the environment, and to life itself. Paolo Freire refers to such reflections as "consciousness as consciousness of consciousness."[1] Ken Wilber makes a distinction between consciousness and form: "the within of things is consciousness, the without of things is form."[2] If you find all of this confusing, rest assured that you are not alone: philosopher Thomas Nagel considers that "The existence of consciousness is both one of the most familiar and one of the most astounding things about the world."[3]

There is no consensus on the definition of consciousness, how it can be measured, where it comes from, where it is located in the body, whether it applies to all matter, and so on. Some equate consciousness with neural activity in the brain and cognition, while others think of it as a fundamental feature of the universe. The lack of a shared understanding of consciousness is not a sufficient reason to dismiss its relevance, especially when it comes to climate change responses. In the 1990s, Danah Zohar and Ian Marshall wrote in *The Quantum Society* that the failure to account for where life and consciousness belong in the universe "leaves human beings with no sense of our place in the scheme of things."[4] If we do not see a place for ourselves in the world or universe, will we feel motivated to even bother engaging with transformations to an equitable and thriving planet?

THE NOOSPHERE

Although there is increasing attention to how social norms and values influence climate-friendly behaviors, the role of consciousness itself is seldom taken seriously in global change research. Climate scientists focus on interactions among the atmosphere, biosphere, hydrosphere, lithosphere, and cryosphere, but they pay little attention to the noosphere. The noosphere refers to the sphere of human consciousness and mental activity, especially regarding its influence on the biosphere and in relation to evolution. The French philosopher and Jesuit priest Pierre Teilhard de Chardin considers the noosphere to represent the interaction of human minds.[5] This "thinking layer" of individual and shared consciousness increasingly includes an awareness of and reflections on our impact on the planet and our role in the universe. Teilhard de Chardin considered the process of self-reflection as critical to the emergence of the noosphere within an evolutionary context: "When for the first time in a living creature instinct perceived itself in its own mirror, the whole world took a pace forward."[6]

Among the scientists who do pay attention to the noosphere, they generally consider it to be a part of the biosphere – for example, in *The Biosphere and Noosphere Reader: Global Environment, Society and Change*, Paul R. Samson and David Pitt define the noosphere as "a process of an increasingly complex intermeshing of cognitive realms within the biosphere – an unfolding of individual and collective ideas, mentalities, aspirations and experience."[7] However, Ken Wilber argues that it is the other way around: that the biosphere is actually part of the noosphere. The reasoning here is that in the same way that the biosphere emerged from and includes the geosphere (or "physiosphere"), the noosphere emerged from (and includes) the organizational structure of the biosphere. This evolutionary distinction is relevant in terms of climate change: "Precisely because the biosphere is a component of the noosphere (and both are thus interdependent), the destruction of the biosphere guarantees the destruction of the noosphere."[8] Undermining the integrity of the biosphere through relentless exploitation, extraction, consumption, and pollution has profound implications for terrestrial and marine species and their habitats, as well as for humans and the ecosystems they depend upon.

Whereas the noosphere has received little attention within the global change research community, the concept and role of the *technosphere* has been discussed and debated in recent years. According to geologist Peter Haff, the technosphere

> includes the world's large-scale energy and resource extraction systems, power generation and transmission systems, communication, transportation, financial and other networks, governments and bureaucracies, cities, factories, farms and myriad other 'built' systems, as well as all the parts of these systems, including computers, windows, tractors, office memos and humans.[9]

The technosphere neither includes nor acknowledges the role of human consciousness, intentionality, or interests in shaping systems, and it trivializes the role of individual and collective agency in transformation processes.[10]

Why has the noosphere received so little attention? Possibly because classical and materialist approaches to science often do not acknowledge that consciousness exists. In the atomistic and reductionist world of classical physics, humans are formed by atoms, molecules, and cells that work together like a brilliant machine, controlled by a central nervous system and a brain made up of billions of electrically excitable neurons that communicate with one another via synapses. Consciousness emerges from these neural activities as integrated information that influences how we think, perceive, and act. Although there are both neural and psychological correlates of consciousness, some scientists consider consciousness itself to be an illusion, along with free will and intentionality.[11]

Wilber argues that physical interpretations of consciousness perpetuate a reductionist approach to science, placing disproportionate significance on particulate matter such as electrons and reinforcing an ego-centered interpretation of reality.[12] His concern is that the subjective experience of consciousness, or what it feels like to "be," can easily become *conflated* or *equated* with material, exterior forms, rather than *related* to them. Like others who interpret consciousness to be fundamental and who challenge the mind-body dualism, Wilber emphasizes that it can and should be studied and validated on its own terms, rather than in terms of physics, whether classical or quantum.[13] Consciousness, from this perspective, is neither meta-physical nor physical. Rather, the physical and metaphysical co-arise as irreducible dimensions of reality.

How we understand and relate to consciousness has implications for education and politics. In *Pedagogy of the Oppressed*, Paolo Freire writes that "Many persons, bound to a mechanistic view of reality, do not perceive that the concrete situation of individuals conditions their consciousness of the world, and that in turn this consciousness conditions their attitudes and their ways of dealing with reality. They think that reality can be transformed mechanistically…"[14] Later, in *Pedagogy of Hope*, Freire emphasizes that political practices based on a mechanistic and deterministic conception of history will never transform dehumanization processes.[15] One could relate this to the non-human world as well, for a mechanistic worldview also renders other species unconscious and dispensable. If consciousness is the experience, awareness, or perception of something, then perhaps social transformations call for new ways of thinking about how and why we matter when it comes to co-creating a world that is equitable and sustainable.

SOCIAL CONSCIOUSNESS

Consciousness has been defined as more than experience. It can also be thought of as the depth and span of the reality that we experience. Developmental perspectives on consciousness recognize that subjective awareness and meaning-making can become deeper and more complex over lifetimes and generations. Such developments typically occur when people or groups are exposed to complex and challenging situations that call for increased self-awareness and a greater capacity to take broader, more complex perspectives on issues.[16] As consciousness develops, information that was previously unnoticed, disassociated, or ignored is integrated into meaning-making.

Not surprisingly, climate change is interpreted and exists differently at different stages of meaning-making.[17] For example, human geographer Gail Hochachka found that Guatemalan coffee farmers with a more traditional meaning-making tended to make sense of climate change mainly in terms of weather or ecological conditions, and as specific impacts, which they then addressed through practical actions. Those with a more "rational" stage of meaning-making had the capacity to perceive the more abstract element of climate change, holding a logical cause-and-effect understanding of it as a problem "to which mechanistic problem-solving may be applied."[18] Those with a more pluralistic meaning-making took a perspective that included an understanding of climate change's multigenerational and cultural impacts, as well as an appreciation for how it contributes to other environmental changes, which can lead to the development of more reflexive and adaptive responses.[19] Such a diversity of meanings exists within families, communities, organizations, nations, and globally, and reminds us of the challenges in establishing a "shared context," as well as to the importance of language and communication.

Changes in meaning-making can lead to different views of the world, different prioritized values, different expressions of agency, and different perspectives on both climate change and the ways to address it. Research on neuroplasticity has confirmed that the brain has the ability to continuously change through a lifetime.[20] Robert Kegan and Lisa Laskow Lahey emphasize that structures of meaning-making can (and often do) change over time. In their research on adult learning, they find that people have the potential to organize meaning-making at whole new levels of complexity, to become more ordered, and to focus their attention and energy on the solutions to complex problems.[21] Kegan and Lahey have identified three systems of adult meaning-making: the socialized mind, the self-authoring mind, and the self-transforming mind. They suggest that "each successive level of mental complexity is formally higher than the preceding one because it can perform the mental functions of the prior level as well as additional functions."[22] This potential has important implications for both social consciousness and social change.

In their research on worldview transformations and the development of social consciousness, Marilyn Schlitz and her colleagues recognize developmental variability regarding the extent to which people are explicitly aware of how culture and society influence them, and in turn how they can impact society, culture, and the environment.[23] This variability ranges from an embedded perspective where consciousness is passively shaped by social, cultural, and biological factors, to perspectives that are self-reflexive, engaged, collaborative, and eventually, resonant. Our relationship to social consciousness depends on the degree of explicit awareness of our interconnections with others.

Perhaps one of the most important aspects of the idea of the Anthropocene is that our *relationship* to the Earth System is changing: humans are becoming self-aware that their actions are influencing global systems. From the perspective of entanglement, when a [whole/part] of the system changes, the system itself has changed. As a thought experiment, consider a world where social consciousness has developed to what Schlitz and her colleagues term "resonant consciousness," a perspective where people sense their essential interrelatedness with others through a field of shared experience and emergence.[24] Resonant social consciousness acknowledges a coherent [I/we] space of oneness, which has been described in philosophy and metaphysics as "pre-space," "the implicate order," "ultimate reality," "unus mundus," or "Source."[25]

Without acknowledging that consciousness and meaning-making can and often do develop over time, humans (and societies) may perceive themselves to be "fixed" in their current roles, identities, and life situations. For many, this means remaining subjected to the whims and wishes of rulers, political leaders, and oppressors who deny their potential to become the subjects or authors of their own lives.[26] For example, within a hierarchical worldview, some people may be treated by those at the top of the hierarchy as objects to be manipulated or controlled, and often they are considered redundant or expendable. The development of consciousness, meaning-making, and agency may be regarded as dangerous to those who have an interest in preserving the status quo. In fact, efforts have been made throughout history, and to this day, to diminish individual and collective agency, often through violence and oppression. Those who abuse power have the capacity to shift their own meaning-making, but seldom the motivation.

When we disregard or overlook the collective, transpersonal nature of human existence, we underestimate our individual and shared potential for generating social change. For example, when it comes to climate change solutions, people are seldom regarded as conscious agents of change. Instead, they are treated as if they were merely "objects to be changed," whether through technical innovations, behavioral interventions, economic incentives, moral persuasion, policies and regulations, or nudging. In denying or undermining the capacity of individuals to develop themselves and their meaning-making over time and to continually engage with the depth and complexity of life, it is quite easy to feel that "most people" do not really matter when it comes to climate change and social change.

QUANTUM CONSCIOUSNESS

Quantum physics has confounded our understanding of reality, leading many to wonder about its relationship to consciousness. Typically, when consciousness meets quantum physics, it is through the idea that consciousness itself is causal and collapses the wave function. This view has achieved particular resonance in what is often referred to as "New Age" thinking. Yet the relationship between consciousness and quantum physics seems to be more complex, and it has been considered and debated from a number of perspectives. In *Quantum Mind and Social Science*, Alexander Wendt challenges the assumption that matter is classical and devoid of mentality.[27] He points to panpsychism, for example, which is the idea that consciousness or mind-like qualities are fundamental to all matter, extending all the way down to sub-atomic particles. The theory of panpsychism traces back to ancient times, and is increasingly being taken seriously by philosophers and some neuroscientists. However, many of these scholars argue that one does not need quantum physics to explain panpsychism.[28] Yet in her book *For Love of Matter: A Contemporary Panpsychism*, ecophilosopher Freya Mathews contends that approaching the universe as "a psychophysical unity" must be consistent with the evidence of science and the requirements of reason, and that this calls for more rationality, not less.[29] If we use reason, she argues, we would eventually ask questions about the nature of reality, which is exactly what quantum physics encourages us to do.

Wendt also draws attention to quantum brain theory, a hypothesis that suggests that the brain, an electrochemical organ, is able to sustain a state of superposition and quantum coherence, and that the collapse of this state translates into the experience of consciousness.[30] Bringing panpsychism and quantum brain theory together, Wendt argues that "at the quantum level, matter is no longer old-fashioned 'matter' – i.e. clearly devoid of mentality – but an ungrounded potentiality to which attributions of mentality make sense."[31] Consciousness – or the world as we experience it – *is* actually the collapse of a wave function of potentiality into a particular experience of reality. That is, consciousness is part of a *quantum* reality.

Wendt's theoretical exploration of the implications of quantum physics for social sciences addresses the mind-body problem in philosophy and the role of consciousness and free will in social life. Rather than being discrete yet interacting individuals, he views humans as "walking wave functions" of potentiality and possibility that intra-act through quantum characteristics.[32] Though Wendt's quantum social theory is speculative, it has started to generate discussions about the role of consciousness within Wendt's field of international relations. It inevitably brings up age-old philosophical and spiritual questions of meaning and mattering.[33] Exploring the interface between science and spirituality from the realm of quantum physics has been contentious, and is often dismissed as unserious, even though physicists such as Niels Bohr, Werner Heisenberg, and others were interested in the deeper meaning of quantum physics.[34]

History tells us that most progressive social changes have been the result of small groups of individuals who see their world in new ways and act from deeper and more inclusive values to challenge existing social and cultural norms. Groups with an expanded sense of social consciousness recognize the rights of previously excluded groups, then work for social and political change. Efforts to end slavery, promote women's rights and civil rights, establish labor rights, support marriage equality, and protect animal rights are not necessarily the *result* of paradigm shifts, but rather the outcomes of people *thinking*, *being*, and *acting* differently. In many cases their actions did eventually shift dominant social and cultural discourses. As Thomas Kuhn notes, paradigm shifts occur when enough people start to live their lives on the basis of that new paradigm, without waiting for the rest of society to make the shift.[35]

IN SHORT....

Consciousness plays an important yet undervalued role in responding to global crises. One of the most profound aspects of climate change, and environmental change in general, is that more people are beginning to consciously perceive themselves to be part of a larger system. When we perceive of ourselves as entangled quantum systems and recognize consciousness and free will as inherent within our being, we can choose to relate differently to ourselves, each other, the environment, and the future. Yet an equitable and thriving world will not just "happen" by adding the adjective "quantum" to social change, or by including a box called "noosphere" to models, frameworks, and strategies. To actualize an equitable and thriving world, we may need a different way of "being in action," including a different understanding of individual and collective agency.

———— REFLECTIONS ————

- Do you think that other species experience some form of consciousness? What about glaciers, clouds, and other forms that influence the climate system?

- If we collectively agreed that everything is to some extent conscious, what difference would it make for the world?

- In what ways do you think your own consciousness and meaning-making have developed over time?

Tina Bjordam
2020

8

AGENCY

AGENCY MATTERS. Transformations to an equitable and sustainable world where all life can thrive will not happen through wishful thinking and hope. On the contrary, agency and action are essential to realizing desired outcomes and impacts. Agency, which can be thought of as conscious actions, intervenes in systems by disrupting or transforming patterns and relationships. From a classical perspective, agency is perceived and experienced as causal, in the sense that an agent acts upon something else, often to produce an intended result. From the perspective of quantum social science, agency is about much more than agents and actions.[1] As Laura Zanotti describes it:

> Agency exists as a way of life, a reiterative activity of opening or foreclosing different possibilities of materialization of matter, not as a relation of push and pull aimed at imposing force on a mass. We are entangled with, constituted and transformed by the very processes we aim to transform.[2]

As an iterative process, agency expresses a *quality* that materializes in the classical world that we perceive and experience every day. This quality of agency can be specifically associated with values that are coherent with – and supportive of – an [I/ we] perspective that promotes thriving for all. This represents a subtle yet powerful shift, and through conscious practices, it can generate responses to global challenges that are "responsible," i.e., based on an awareness and respect for connections and nonlocal entanglements. In contrast to the individualistic and fragmented dualisms of our classical world, agency in a quantum world acknowledges that [I/we] are [whole/parts] in a dynamic and relational process of being and becoming.

A BELIEF IN AGENCY

Both agential realism and participatory realism suggest that we are part of the phenomena that we are measuring, yet our sense of agency is closely tied to the beliefs that we hold about the world and our relationship to it. Beliefs are particularly relevant to gender rights, human rights, labor rights, children's rights, animal rights, and our understanding of rights in general. They also influence ideologies, or "isms," including racism. Author Isabel Wilkerson makes this point effectively in *Caste: The Origins of our Discontents*:

> When people have lived with assumptions long enough, passed down through the generations as incontrovertible fact, they are accepted as the truths of physics, no longer needing even to be spoken. They are as true and as unremarkable as water flowing through rivers or the air that we breathe.[3]

As discussed earlier, there is good reason to pay attention to the beliefs and assumptions that we hold, and to reflect not only on how they are influenced by the wider social and cultural context, but also on how they influence our sense of agency.

The role of agents and their subjectivity (i.e., the so-called "interior" world of beliefs, emotions, identities, perceptions, etc.) in consciously shaping physical and social worlds has long been a source of tension between realists and idealists. Very generally, realists maintain that there is a world out there that is separate and distinct from humans, and as such they may favor radical empiricism, which holds that reality can only be attributable to that which one can observe. Idealists recognize that ideas, perceptions, and understandings shape not only the way we experience this world, but also the world that we experience. Transcendental idealism maintains that reality is only accessible to people as an individual or social construction.[4] These examples highlight extreme views of what is real; however, "in reality" there are also perspectives and metatheories that transcend realism and idealism, including critical realism and integral theory.[5]

Each of these philosophical perspectives takes a particular view of the role of subjective beliefs and human agency in relation to the external world. For instance, let's consider a purely realist perspective, where beliefs are considered a product of our knowledge about the world out there – a world that exists and can be observed. In this case, "we believe what we see." As an example, measurements of ice velocity, subglacial water pressure and meteorological variables from the western margin of the Greenland ice sheet tell us that the ice sheet is more vulnerable than previously believed.[6] This conclusion is based on measurements and data, and the changing conditions of the Greenland ice sheet are alarming for many reasons. Now let's shift to a purely idealist perspective, where reality is seen as something that is filtered through our subjective experiences, emotions, cultural and social norms, beliefs, assumptions, and expectations. From this perspective, "we see what we believe." If one does not believe that human activities are contributing to climate change, the observed melting of the Greenland ice sheet may be interpreted as part of normal variability, or as part of a gradual recovery of global temperatures from the Little Ice Age.[7]

Although realism, naturalism, physicalism, positivism, idealism, interpretivism, and many other "isms" have created interesting debates within the philosophy of science, these perspectives matter; they have consequences for policies and practices, particularly in relation to climate change. For example, when Irene Lorenzoni and Mike Hulme showed projected future climate scenarios to people in both the United Kingdom and Italy, they found that the scenarios did not really influence people's attitudes about the future.[8] Whether or not people believed them to be credible depended on their prior beliefs and their trust in the science of climate change. The results of their study support other research showing that preexisting beliefs shape perceptions and attitudes related to climate change.[9] Such perceptions influence the actions that we take – or do not take – in response to global challenges. In other words, they influence whether and how we view and express our agency within an entangled quantum system.

AGENCY IS NOT NEUTRAL

There is no such thing as a value-neutral response to climate change, and not every response will have an equally positive impact on the whole. In fact, some responses, such as geoengineering of the oceans or atmosphere, could have negative consequences for many people and species.[10] As such, it is important to make the underlying values and intentions behind climate responses transparent. Attention to underlying or hidden values is especially critical in an era of "big data," where algorithms are used to influence decisions and investments. Cathy O'Neil points out that "the same models that inflict damage can be used to benefit humanity, and that the heart of the problem is almost always the objective."[11] The objective she refers to relates to the intention or purpose for which the data will be used, and these are closely linked to the values that underlie them. For example, in the so-called "carbon economy," carbon trading schemes are tied to pricing algorithms designed to maximize economic efficiency, rather than to reduce greenhouse gas emissions.[12]

Many actions today are expressions of agency based on exclusionary values that favor only a fraction of society. Political economies and financial instruments are currently concentrating wealth in the hands of a few, and rather than creating global peace and prosperity, this wealth has often been used to finance environmental destruction, ill health, and war.[13] The hierarchies of domination and exploitation that some consider necessary to maintain the status quo tend to be tied to worldviews that retain an "us versus them" and "humans versus nature" approach to meaning-making.

Valuing wealth as an entitlement for the privileged often comes at the expense of well-being for all people, species, ecosystems, and the planet. Not surprisingly, much of society is currently organized to maintain the myth of separation and difference. Paolo Freire points out that the very idea of unity is considered to be a dangerous concept for those seeking to uphold the status quo.[14] Consequently, actions that promote unity, and particularly "unity through diversity," can be perceived or experienced as threatening by some. When people have been led to believe that a fragmented world is the only possibility, and that anything else is simply an illusion or delusional, it is unsurprising that few people dare to speak out and work for so-called "impossible" outcomes that are based on values inherent to [I/we] and [whole/parts].

The sharp contrast between a fragmented world characterized by structural inequality and a deteriorating environment, and a world that is both diverse and thriving, can be a powerful driver of social change. In fact, concerns about climate change are motivated by a recognition that its severe, widespread, and irreversible impacts will affect all [whole/parts] of the planet. This recognition has mobilized millions of people to take action, as evidenced through the growing number of sustainability initiatives around the world. Nevertheless, calls for radical transformations are typically met with resistance, especially by those with vested interests in current systems, and this includes many of us. Having a strategy to shift systems and cultures is thus essential, but as leadership development practitioner Monica Sharma reminds us, "do not expect people to clap for you when you rattle the system."[15] This is why leadership is important to quantum social change.

Leaders may both enable and resist social change. Within the field of international relations, Alexander Wendt discusses the role of leaders in collapsing the state's wave function into one among many potential outcomes. In relinquishing power to a leader with the authority to decide on behalf of the collective, each person in an entangled system momentarily gives up their choice on how to respond. According to Wendt, there are two important implications of transferring agency to a "leader." Firstly, he recognizes that "the intentions and character of leaders are crucial in determining which policies are realized. Even in highly constrained situations, small differences in leaders can make big differences in what actually happens."[16] This is evident in the world today, where elected leaders of some of the largest countries have enacted policies that fracture society and accelerate environmental change, instead of addressing the drivers. Secondly, Wendt argues that when a leader collapses a state's potentialities into an actual choice, it has non-local consequences.[17] The ripple effects of one leader's words, decisions, and policies regarding climate change and the environment often have significant impacts across time and space. When leaders fail to stand for values that apply to everyone, or show few signs of integrity, the results can be destructive and even disastrous. Leadership, character, and intentions matter much more than we think. Although newly-elected leaders can redirect the course of action, quantum social change recognizes that everyone has the capacity to shift systems and cultures, no matter what their position.

THE QUALITY OF AGENCY

It is thus not only a belief in agency that matters, but also the *quality* of agency that contributes to quantum social change. This quality is related to the values that we stand for, both for ourselves and others. Henry Stapp emphasizes that values are ultimately tied to the beliefs about one's relationship to the rest of the universe, and "what we value depends on what we believe, and what we believe is strongly influenced by science."[18] This may be true for Stapp, who is a physicist, but many people's beliefs are currently not influenced by science. Values, however, can create a strong bridge between science and culture. In a study of climate responses among Andean farmers, human geographer Morgan Scoville-Simonds found that values and beliefs were embedded within "entangled narratives" of both climatic and cultural change.[19]

Relating to the world as a quantum system calls for actions that are sourced from values such as integrity or wholeness. Referring to the quantum Zeno effect, psychotherapist Ton Baggerman emphasizes that awareness of our values can play a significant role in how we co-create our reality.[20] In each moment we are both passively and actively co-creating structures and systems that support or inhibit a thriving world. Agency – or the actions that we take – provides us with feedback, information, and knowledge that we can use to update our beliefs and change our "bets" on the future. Sustainability is no longer a normative goal, but a potential that can be realized right here and now. As political scientist Karin Fierke writes:

> The physical basis of our conceptualizations that is provided by quantum physics transforms ethics from a purely normative enterprise focused on what should be – which goes against the grain of what is – to an enterprise focused on potentialities. In turn, this opens a space for agency.[21]

Values that express integrity, or the state of being whole, undivided, and coherent, are powerful in generating social change. Expanding the definition of "value" to include more than economic interests makes it possible to embed relational qualities, such as equity and justice, within all aspects of sustainability research, policy, and practices. For example, human geographer Milda Nordbø Rosenberg found that values of togetherness, care, dignity, and faith transformed coffee production systems in Burundi, with profound impacts on lives and livelihoods.[22] In *Radical Transfor-mational Leadership: Strategic Action for Change Agents,* Monica Sharma refers to values that are inherent to all humans (and arguably to non-humans as well) as universal values.[23] Many of these universal values have been directly acknowledged in the 1948 *Universal Declaration of Human Rights,* which recognizes the inherent dignity and equal rights of all members of the human family as the foundation of freedom, justice, and peace in the world.[24] Expressing agency based on these deep and innate values allows us to access a shared context and a space of entanglement that holds the potential to generate non-local change.

AGENCY AS RESPONSIBILITY

When it comes to agency, there is an enormous difference between talking about values and embodying them. Monica Sharma writes that we create impact when we embody and express our inner space of oneness as we act from moment-to-moment.[25] When actions of individuals consistently embody universal values of equity, dignity, fairness, and compassion for all, new patterns can be created within communities, cities, nations, and global institutions. Responses that are grounded in values that apply to [whole/parts] dissolve discrepancies between short- and long-term goals, and between the well-being of current and future generations. Whether in reference to the development of renewable energy resources, regenerative agriculture, a circular economy, a sharing economy, or a respect for each other and all species, the actions that we take now create nonlocal effects across space and time. The climate system is responsive to human actions, which means that with awareness, attention, and intention, every instant has the potential to contribute to sustainability, not just as a normative goal, but as a way of being.

To deliberately transform political, economic, social, technological, and cultural systems and structures requires the perception and activation of free will. Henry Stapp describes free will as "the capacity of mental intent to influence physical behavior."[26] Alexander Wendt also identifies "will" as being critical to agency: "Will is the essence of agency, a power to animate and move the body – and the mind, in the form of attention – from the essentially passive stance of Cognition to active, purposeful engagement with the world."[27] Whenever we willfully intend to generate sustainability, social justice, and peace, we direct both intention and attention towards this. Success often involves breaking with habitual patterns of thoughts and beliefs. For Laura Zanotti, this includes "the [materialist] conviction that humans can master and shape social transformations according to the design of planning rationalities."[28] Such rationalities, though well-intentioned, are often based on dualistic, deterministic, and fragmented views of the world.

The ability to transform systems calls for a new way of conceptualizing our relationship to political change. Political agency, as Zanotti describes it, "is rooted in ambiguity and performativity – i.e., on the making and remaking of meaning, subjects, power and political spaces in the context of intra-active relations."[29] Her practice-based approach to political agency draws on Karen Barad's concept of "agential realism," which recognizes humans as "agencies of observation" who participate within nature and thus are constitutive of reality.[30] Barad considers space, time, and matter to be dynamically generated through intra-actions; she considers *phenomena* to be the smallest units of relation that come to "matter" through ongoing and entangled intra-activity:

> Phenomena are constitutive of reality. Parts of the world are always intra-acting with other parts of the world, and it is through the specific intra-actions that a differential sense of being—with boundaries, properties, cause, and effect—is enacted in the ongoing ebb and flow of agency.[31]

Barad's interpretation emphasizes that agency is not an attribute that someone or something has, but an *enactment*. Indeed, for Barad, matter is "not a thing but a doing, a congealing of agency" through which phenomena are constantly being materialized and come to matter.[32] She stresses that both human and non-human intra-actions matter in reconfiguring the world. However, in everyday life, we tend to perceive agential "separability," i.e., the sense of a difference between subjects and objects. From an agential realist perspective, such separation is always enacted within a particular context or phenomenon.

Humans, with a capacity and potential to be reflective and "conscious of consciousness," have a particular "response-ability" to respond in ways that support [I/we] and [whole/parts]. As Barad puts it, "we are always already responsible to others with whom or which we are entangled."[33] This includes responsibility for the norms, rules, standards, regulations, institutions, incentives, and power relations that influence how society is organized and who benefits. Does "who benefits" include all groups, species, ecosystems, generations, and processes that enable life to flourish? In accepting our response-ability, we can engage with conscious practices that shift cultures and systems to support a thriving planet.

IN SHORT

Transformations to an equitable and thriving world call for more than hope and optimism. They call for actions and interventions that challenge outdated systems by generating new ones that are more equitable, diverse, and inclusive. Karen Barad's concept of agential realism emphasizes that we are constantly intra-acting with the universe, and our intra-actions matter because each one reconfigures the world.[34] Through our entangled intra-actions, we are mattering in every moment. But it's not just the expression of agency that matters. Rather, it is the quality of agency that we are interested in; a quality that recognizes oneness and is expressed through values inherent to the whole, such as equity, diversity, and compassion. When these values are at the heart of individual agency, collective agency, and political agency, it is possible to generate new, fractal-like patterns that replicate across scales, in every moment.

—— REFLECTIONS ——

- When it comes to the quality of agency, what values are most important to you?

- Have you ever stood up for others in the face of injustice? If so, how did that feel? What are some of the obstacles that might inhibit you from taking a stand?

- What are some ways that [I/we] can express political agency to generate an equitable and thriving world?

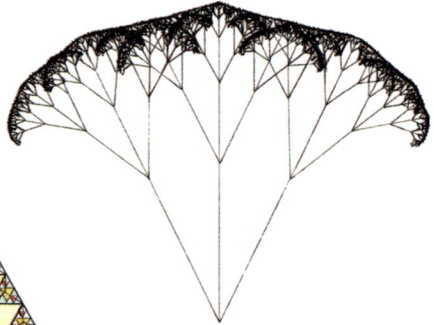

9

FRACTALS

FRACTALS MATTER. A fractal is a self-similar pattern that repeats itself, regardless of scale, and it is an example of what David Bohm and F. David Peat refer to as a "generative order."[1] Fractals are visible throughout nature, with examples ranging from beehives to tree branches, leaves, dragonfly eyes, and river systems. Fractals in geometry and algebra are formed by calculating an equation or repeating a process over and over. Examples include the Mandelbrot Set and the Sierpinski Triangle, which produce self-similar patterns that can be magnified or reduced to any scale, including subatomic scales. Fractals also offer us a different way of thinking about scaling social change.[2]

Fractals are not to be confused with fragments. Bohm writes about the deep and pervasive roots of our fragmentary way of thinking, looking, and acting, which has implications in every aspect of human life.[3] For example, fragmentation often means that diversity, rather than being appreciated and celebrated, is interpreted as difference and "otherness." Values such as greed, cynicism, or elitism, when embodied and replicated across scales, can be effective in maintaining a status quo that depends on profit maximization, labor exploitation, environmental degradation, and so on. This often results in a polarized and conflicted society based on dominator hierarchies that suppress rather than emancipate human potential. Patterns that externalize other people, species, or nature in general produce and reproduce fragmented conditions that are contributing to a high risk of abrupt and irreversible changes, or "tipping points" in the Earth System.[4]

Quantum physics supports what Bohm calls a "whole movement" that is unbroken and undivided.[5] Within this whole movement, every aspect flows into and merges with every other aspect. To move from a fragmented world that feels fractured and broken towards a whole movement that is healthy and coherent, Bohm and Peat suggest we focus on a generative order that is based on a repetition of patterns and principles: "This order is primarily concerned not with the outward side of development, and evolution in a sequence of successions, but with a deeper and more inward order out of which the manifest form of things can emerge creatively."[6] This is closely related to Bohm's idea of the implicate order, which refers to a deeper level of reality influencing the explicate order or the external world that we actually perceive.[7] As a generative order, fractals represent a potentially powerful approach to scaling social change.

SOCIAL FRACTALS

If we are interested in social change, we need to consider not just fractals in nature or mathematics, but also social fractals. These can be defined as self-similar patterns generated by individuals and groups from moment to moment, through conscious intent and agency. Although often considered no more than a metaphor, social fractals are real in that they can transform relationships and generate new patterns and structures in society. For example, if our actions are consistent with values that apply to the whole and resonate through language, meaning-making, and values, they will replicate, generating new patterns that reflect these qualities at all scales. When Greta Thunberg went on strike for climate action based on values of integrity and a respect for nature, she activated political agency in many other young people around the world, leading to the Fridays for Future school strikes and legal challenges to inadequate national climate policies. In *The Politics of Waking Up,* activist and psychotherapist Indra Adnan defines fractal politics as the pattern of relationships at the heart of complex systems: "a fractal politics must shift the focus away from the vertical power of the state and look for the local and municipal energy that can connect human flourishing with planetary flourishing."[8] Adnan points to community institutions as a particularly powerful and influential fractal, recognizing that it is the quality of our relationships that creates new patterns and possibilities. This includes relationships with both sentient and non-sentient beings.

In describing the process of creating an ecological civilization, sustainability theorist and activist Paul Downton stresses the importance of conscious, systemic cultural change, and he points out that "the social framework and values that underlie the making of human settlement are critical to the creation and definition of an ecological city."[9] He describes cultural fractals as "a living system of human relationships that displays the essential characteristics of the larger culture of which it is a part."[10] In *Integral City*, Marilyn Hamilton discusses the fractal relationships of micro, meso, and macro human systems and how these fractal patterns "reveal that the health of the city is deeply embedded in the patterns or rules that contribute to the health of individual families, the team, the organization, the neighborhood, city hall, nation and the world."[11] An urban fractal thus contains the essential characteristics of the larger urban whole of which it is a part – or rather, a [whole/part].

Institutions may change over time and structures may be reformed, updated, or repackaged to appear as new and more progressive, but if their underlying values do not embody or promote thriving for all, at some scale they will generate fragments rather than fractal patterns. Fragmented responses are unsustainable; even when redesigned or improved, the patterns are likely to perpetuate divisions within and across scales, rather than alignment and synergies. Yet, it is possible to consciously and purposively generate social fractals based on values that are consistent with a thriving world. Monica Sharma describes a fractal approach to shifting systems and cultures at scale, recognizing that each idea, initiative, or endeavor can be designed with the same characteristics desired for the whole.[12] By basing our intra-actions on the values we wish to see replicated in the world, we can generate self-similar patterns that apply to everyone, and to all situations, whether at home, at work, in a community or organization, or through political parties and social activism. These fractal patterns, in turn, can generate results at all scales.

SOCIAL CHANGE AS A FRACTAL PROCESS

Social change can happen rapidly when [I/we] disrupt inequitable and unsustainable patterns and generate new ones aligned with values that are inherent to entangled [whole/parts]. Still, many people question whether we still have time to transform in an equitable and sustainable manner, given the climate changes already underway and the widespread loss of biodiversity. Interestingly, quantum physics challenges the dominant perspective of time as uni-directional and linear, upsetting our perceptions of the relationship between past, present, and future. It also introduces ideas such as time entanglement and non-local experiences in time.

In *The Order of Time*, Carlo Rovelli suggests that the intrinsic quantum indeterminacy of things results in a blurred vision of the world that dissolves the distinction between the past and the future.[13] Similarly, Alexander Wendt draws on philosophical perspectives about a "generalized quantum theory that describes an underlying reality in which there is neither time nor a distinction between mind and matter," emphasizing that collapse of the wave function does not takes place in time, but "*it is in collapse that 'Now' emerges*, and with it the distinction between past and future."[14] Wendt's quantum social theory is based on the idea that the future can influence the present, creating a non-local form of causality.[15] The idea that the future influences the present has long been acknowledged in many cultures. For example, the Onondaga Nation recognizes the wisdom of the Peacemaker, a divine messenger whose words of unity and harmony not only emphasized the relationship between individual health and societal peace, but also the importance of making decisions on behalf of the next seven generations, to ensure they will enjoy a good life, and also to bring peace in the present time.[16]

Our experience, from a subjective perspective, is always taking place *within* a process of change. Or as Rovelli writes, "In order to understand time, it is not enough to think of it from outside: it is necessary to understand that we, in every moment of our experience, are situated *within* time."[17] This is not just a philosophical musing; in every instant we are thinking thoughts and making decisions that are influenced by the environment, and these influence the sources and sinks of greenhouse gases, which will have consequences for Earth System processes for centuries and generations to come. This invites us to consider the concept of quantum fractals. If [I/we] are [whole/parts] in an entangled universe, our being and doing in the moment actually matters for all of us, and in a manner that transcends classical notions of space and time.

An agential, participatory approach recognizes that both humans and non-humans play a role in social change processes. Barad's agential realism involves a shift in perspective from a world of substance, matter, and things to a universe of processes and phenomena; her understanding of space, time and "mattering" seems to align with the process philosophy of Alfred North Whitehead, a mathematician and philosopher writing in the early 20th century. Whitehead's speculative philosophy, as described in his essay on *Process and Reality*, represents a radically different approach to the concept of matter, challenging the idea that matter exists all.[18] He suggests that nature is a process, or as geographer Tom Roberts describes it, "Things are, only insofar as they are taking place."[19] Whitehead's process philosophy includes the idea that "the subjective and the objective [are] both joined together actively in every event to create greater complexity and possibility."[20]

Whitehead refers to the smallest instance of process as an "actual entity" or an "actual occasion" which is a singularity of the present that is influenced both by reverberations from the past and anticipation of the future: These are "the final real things of which the world is made up."[21] Although the patterns and relationships generated by actual occasions create the reality that we perceive and experience, in Whitehead's process philosophy, the occasions themselves are considered to be causally independent of one another. In Whitehead's view, the linear sequence of time that we perceive to be "real" is actually a continual process of occurrences. According to process philosophy, matter is an *event* that is influenced by or connected with previous events: when we consider ourselves as a process, there is no "being" without "becoming."

Process philosophy and quantum physics both emphasize the relationship between discrete and continuous phenomena.[22] This is explained by Henry Stapp in terms of atomic actualities and continuous potentialities, where the world of fixed and settled facts grows via a sequence of actual occasions.[23] In Whitehead's words, "Immediacy is the realization of the potentialities of the past, and is the storehouse of the potentialities for the future."[24] Drawing on Whitehead's process philosophy, physicist John A. Jungerman also interprets the world at its most elementary level to be in the realm of events or phenomena, not substances.[25] This is based on the idea that solid matter is, to an astonishing degree, empty space "filled with an unimaginable number of events – not only those from spontaneous particle pairs but also from the virtual photons, gluons, bosons, and gravitons that are the force carriers."[26] The unimaginable number that Jungerman refers to is on the scale of a trillion trillion. Compare this to the capacities of the best high-speed cameras, which can capture ten trillion frames-per-second. Although temporal resolutions of a quadrillion frames per second are not far off, the difference between our theories and perceptual capacities remains vast.[27]

Since the collapse of a wave function through a sequence of actual occasions occurs over an incredibly short time scale, it is easy to understand why we perceive of matter from a classical perspective. Reality feels very "solid," and we experience consistency over what seems to be a continuous flow of time. However, quantum social change emphasizes our potential to consciously disrupt habitual patterns and generate new ones. In fact, Stapp argues that successful living demands the ongoing generation of value-based intentional actions in every moment.[28] R. Buckminster Fuller phrased this potential quite poetically: "I am not a thing – a noun. I seem to be a verb, an evolutionary process – an integral function of the universe."[29] This is another way of saying "we matter in the moment."

SCALING TRANSFORMATIVE CHANGE

How do we put these insights on processes and fractal approaches to scaling change into practice? Many people are overwhelmed by a combination of the scale of global crises and a limited perception of their personal agency. The narrative of the "hero's journey" emphasizes the individualized nature of human struggles, including the challenges of integrating conscious and unconscious aspects of the psyche.[30] At the same time, it's clear that large-scale change requires collective action. Interestingly, there is a tendency to approach social change in terms of individual change *versus* collective change, bottom-up change *versus* top-down change, or local change *versus* global change. Such dualisms overlook the importance of a [both/and] perspective, and can lead to partial and polarizing responses. Within a classical paradigm, approaches to scaling change tend to include efforts to scale *up* by aggregating to higher levels of authority or to scale *out* through replications based on certain standards or the sharing of knowledge and best practices, often with little attention to context. However, the importance of cultures and values has been recognized, and expressed by Michele-Lee Moore and colleagues through the concept of scaling *deep*.[31] Such approaches to scaling change are important, yet they rarely recognize nonlocal entanglement among [whole/parts], or consider [I/we] to be fractals of change.

Scaling is not a linear process. The "Powers of Ten" framework (Figure 2) developed by Mark McCaffrey, Avit Bhowmik, and colleagues, examines the relative population concentration from scales ranging from the individual (10^0) to families (10^1), personal networks (10^2), villages (10^3), communities (10^4), metacommunities (10^5), urban areas (10^6) and so on, all the way up to the global scale (10^{10}).[32] This exponential view is powerful, in that it illustrates how [whole/parts] are embedded in different modes of social organization. McCaffrey and Bhowmik's team explores the related targets and indicators for successful engagement at each level, as well as possible interventions and barriers. They suggest that there is a "sweet spot" in the middle where local and global converge, and where action can have the greatest impact, and that this lies in the range of communities (10^4) and metacommunities (10^5), i.e., in the range of ten thousand to one hundred thousand people.[33] Importantly, this range corresponds with the scale that Indra Adnan suggests is the most effective for the practice of fractal politics.[34]

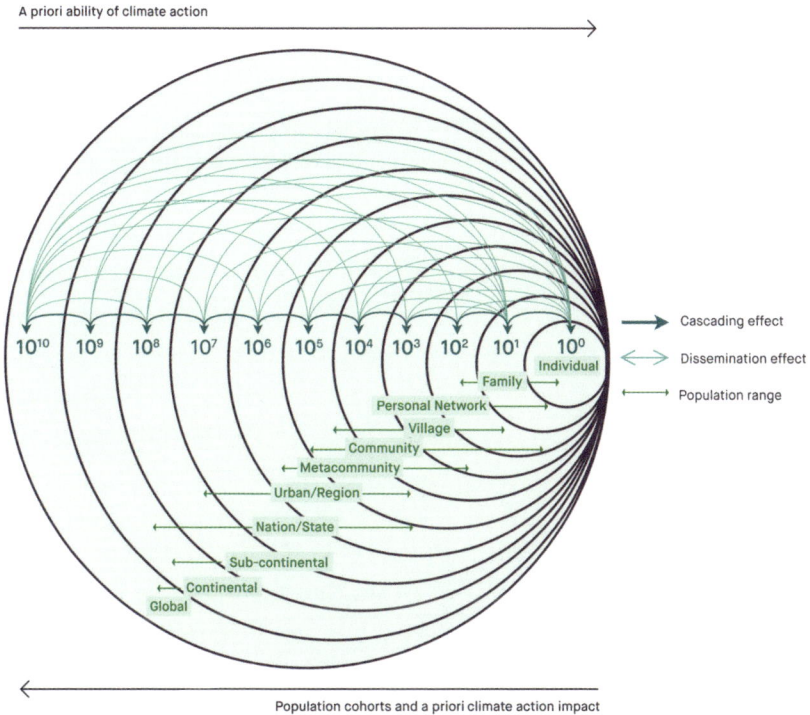

Figure 2: Powers of 10 Framework (based on Bhowmik et al. 2020).

This brings us back to the role of individuals in scaling nonlinear social change. Can one individual (i.e., 10^0) make a difference at scale? Recall that unlike natural or mathematical fractals, which generate value-neutral patterns, human and social fractals embed values that replicate a particular *quality* at all scales. Values are thus woven into the patterns that structure our relationships with each other and the environment, generating fractal patterns that both resonate and replicate at all scales. These fractals are context-specific and emerge through conscious agency and intra-actions. Emergence, as Alexander Wendt sees it, is an outcome of our intra-actions: "This view finds emergence not in vertical relationships between levels, but in holistic horizontal ones among agents, whose states are constituted by non-local entanglements mediated by language."[35] He recognizes both social structures and agents as superpositions of potentiality:

> Their superposed states are co-emergent, and if they become real realities they do so together in localized practices, which themselves are emergent from the dynamic processes of wave function collapse.[36]

Political scientists Karin Fierke and Francisco Antonio-Alfonso provide a nice analogy that illustrates how values embedded in fractals actually scale: "Much as the potter molds clay into a finished shape, particular patterns of behavior lead to particular results that reverberate across time and space, and in this respect individuals and societies create themselves through their moral choices while also creating their futures."[37]

As discussed throughout this book, quantum social change is based on the idea that we are entangled through our intra-actions in a quantum system, including through shared meaning-making, values, language, and stories. In his book, *On Dialogue*, David Bohm writes that our words generate a stream of meaning that seem to flow "among and through us and between us."[38] He emphasizes that "this shared meaning is the 'glue' or 'cement' that holds people and societies together."[39] This is not, however, the classical glue that we are familiar with; it is a relational bond that connects [whole/parts] and [I/we] across time and space.

Fractals allow us to work simultaneously at multiple scales; every action holds the potential to generate patterns that replicate nonlocally through language, shared meaning, and values. As Downton puts it, "Effective change agents must be microcosms of the change they seek to bring about."[40] If we think of emergence as both a horizontal and a vertical phenomenon, every person really does matter. In fact, every conversation and every intra-action has potential and can be significant. Whether we use the language of ripples, waves, or resonance, the point is that our agency can generate quantum fractals that replicate at all scales.

The recent discovery of quantum fractal patterns in electrons suggests that this idea may go beyond a metaphor. As described by Arkadiusz Jadczyk:

> As of today, quantum fractals are just images and algorithms. They are pretty. But in the future, I am sure, we will find them all around us. And not only in the highly specialized realm of quantum microphysics, but also in macrocosmic light patterns, and also in the extensions of the quantum formalism that regulate social and psychic phenomena, processes in the brain, thought formation, etc.[41]

From the perspective of quantum fractals, every person who takes actions based on universal values such as equity, dignity, compassion, and integrity affects all scales. Connecting to others from an [I/we] space transcends separation and fragmentation. Monica Sharma writes that embodying values and acting with integrity enables and requires us to speak up, speak out, and redress violations to the environment and to human rights.[42] This involves a recognition of rights, but also of connections, coherence, and wholeness. Challenging the systems and practices that reproduce unequal and unsustainable outcomes, as well as violence, involves not just individual and collective agency, but political agency, or the capacity to consciously generate new structures and systems that benefit humans and non-humans alike. This entangled [I/we] space represents a powerful "sweet spot" for social change. The question is, how do we access this space, right here and now?

In the classical world, paradigm shifts can take a long time. In fact, there is a widely-held belief that we have to wait for older generations to die before new thought patterns replace the old ones. Yet in an entangled quantum world, it is possible to inhabit and experience new paradigms or thought patterns, right here and now. In alignment with process philosophy, Bohm and Peat emphasize that knowledge is "not something rigid and fixed that accumulates indefinitely in a steady way but it is a continual process of change."[43] It takes no time at all to shift from an "us versus them" perspective to an [I/we] perspective, and the potential for an equitable and thriving world exists in every moment. When we consciously create "quantum fractals" based on values that apply to every living being, we contribute to wholeness rather than fragmentation. The more that people generate such "quantum fractal" patterns, the sooner we will notice structural and material changes in society. To realize these transformations "in time," we have to be conscious of the way we relate to each other, to nature, and to change itself. Karen Barad reminds us, "The world and its possibilities for becoming are remade with each moment."[44]

In short...

Quantum social change involves generating self-similar patterns that repeat at all scales. Every action represents an opportunity to both individually and collectively influence the whole. This represents a different way of relating to social change, grounded in universal values and our way of being. It is aligned with a process-oriented approach to change which recognizes that the future is conditional on how we are and what we do right now. Recognizing that every person matters in every moment, quantum social change acknowledges that one small change can have significant nonlocal impacts when it reflects values that apply to the whole. In the classical world, paradigm shifts can take a long time, but in a quantum entangled world, we activate the new paradigm when we consciously generate "quantum fractals" that contribute to wholeness and diversity, rather than fragmentation. This is where you come in, and why you matter.

—— REFLECTIONS ——

- Can you think of examples where the values expressed by individuals, groups, organizations, or governments have created fragments, rather than fractals?

- If you were to consider yourself a fractal of change, what value(s) would you like to replicate at all scales to generate an equitable and thriving world?

- In relation to the Powers of 10 Framework, what are your primary spheres of influence?

10

YOU

YOU MATTER. In fact, as a [whole/part] of an entangled quantum system, you matter much more than you think. When you and [I/we] relate to ourselves, each other, the environment, and the future from a quantum perspective, it becomes clear that we matter in every moment. Agency is an entangled, intra-acting phenomenon, and the outcomes of our intentions and actions affect us all, whether we are aware of it or not. Fortunately, we have the capacity to consciously intra-act and in doing so, can transform our individual *and* shared realities. Quantum social change involves consciously contributing to an equitable and thriving world in every moment, based on values that acknowledge [I/we], [whole/parts], and [both/and] relationships. It is the practice of sowing the seeds of a thriving future for all by allowing ourselves to embody that future in our thoughts, relationships, and actions. When we embrace our complementary identities, as *both* individuals *and* collectives, we can generate fractal patterns that ripple, resonate, and replicate at all scales.

MAKING A DIFFERENCE

The metaphors and meanings associated with quantum social change will do nothing unless they are coupled to frameworks and strategies that have an impact. One practical, well-tested approach that generates results is Monica Sharma's conscious full spectrum response, which is described in *Radical Transformational Leadership*.[1] This framework has been adapted to climate change and sustainability through the "Three Spheres of Transformation," a model that represents interactions among the practical, political, and personal spheres (see Figure 3).[2] Shifting from partial responses to "full spectrum" responses acknowledges the significance of [whole/parts], [I/we], and [both/and] perspectives in transformation processes. The Three Spheres of Transformation represents a fractal approach to social change; it recognizes that transformations involve integrated responses that simultaneously address multiple dimensions of change.

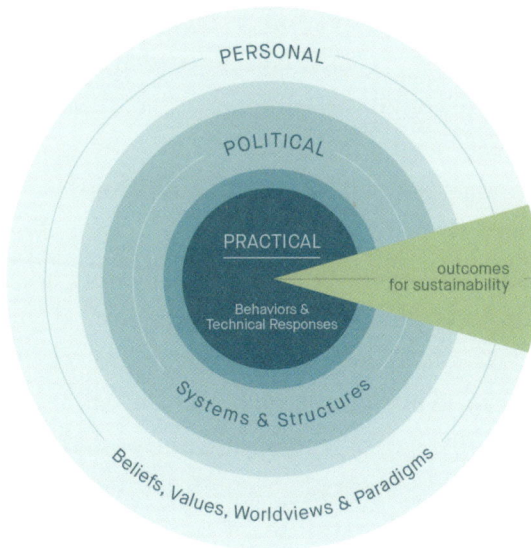

Figure 3: The Three Spheres of Transformation (Source: O'Brien and Sygna 2013).

Most sustainability initiatives currently focus on the practical sphere, which includes the technical and behavioral responses that are directly associated with desired outcomes. For example, solar panels, bicycles, windmills, electric vehicles, reduced food waste, plant-based diets, and other solutions can help to lower greenhouse gas emissions and meet the goals established in the Paris Agreement on climate change. However, the success of these changes is often facilitated or hindered by systems and structures, including social and cultural norms, rules, regulations, institutions, and infrastructure that influence how societies are collectively organized, what is prioritized, how resources are managed and distributed, and so forth. These are referred to as the political sphere, recognizing that they are influenced by decisions and actions taken by or on behalf of the collective.

Mobilization and collaboration in the political sphere can spark collective action, whether in relation to removing subsidies for fossil fuels, creating bike lanes, establishing recycling systems, implementing effective early warning systems, designing robust health care systems, promoting equitable economic policies, institutionalizing justice and a respect for human rights, and other structural changes. Actions in the political sphere of transformation are often influenced by power relations and vested interests, which may lead to situations of conflict. As a result, political changes are often easier said than done, particularly in polarized contexts, or where power hierarchies deliberately limit or undermine the capacity of all people to have a voice and to thrive. However, such contexts can also mobilize social movements and collective action aimed at achieving practical, measurable results. The impact of such movements and actions depends on the capacity to strategically design, implement, and navigate systems change.

This is why the personal sphere of transformation is significant. Actions and interventions in the practical and political spheres are influenced by our individual and shared beliefs, values, worldviews, and paradigms. These shape how individuals and groups relate to others, engage with systems, view leadership, understand causality, and so on. They also influence the goals that are prioritized. When systems and strategies are designed based on universal values such as equity, compassion, dignity, and integrity, the results reflect those qualities. Yet, the personal sphere of transformation still tends to be ignored in many sustainable development initiatives, and when it is addressed, it is most often through efforts to *change* the beliefs, values, worldviews, and paradigms of "others," rather than to create conditions for everyone to realize their own potential based on values that resonate and replicate at all scales. This cannot be stressed enough: a fractal approach to transformative change is not about changing others, rather it is about showing up and "mattering" in ways that transform the cultures and systems that are perpetuating inequitable and unsustainable practices, to create results that enable all life to thrive. Transformative change often starts with self-inquiry: What is important to me? What do I stand for, both for myself and others? What results do I want to see in this world? And what systems and cultures need to shift to realize these?

The Three Spheres of Transformation is a fractal that places *you* at the center of quantum social change. In contrast to classical approaches to social change, quantum social change is based on the idea that individuals are "phenomena," and that [I/we] can generate entangled fractal patterns through our thoughts, ideas, words, metaphors, decisions, conversations, actions, and agency. Returning to the "Powers of 10" framework developed by Mark McCaffrey, Avit Bhowmik, and colleagues,[3] which was described in the previous chapter, "you" (10^0, depicted as P0 in Figure 4) are located at the center of exponential change. Self-awareness and reflexivity, combined with attention and intention, can generate fractal patterns that resonate constructively with family and friends (P1), personal networks (P2), communities (P3), villages or neighborhoods (P4), metacommunities (P5), and so on, extending to urban, national, regional, continental, and global scales.

The point is that you, me, and every other person are nonlocally entangled with others, thus we are always influencing [whole/parts]. Our "mattering" transcends time: the scale of (P11) corresponds to all *Homo sapiens* who have ever lived, represented as our shared humanity.[4] Though our work in any moment may focus on a single scale or "sphere of influence," our impact extends across all scales, entangled through the values embedded in our language and meaning-making. Fragmentation occurs when the entanglement of [I/we] and [whole/parts] is collapsed into a reality that is based on exclusion, exploitation, oppression, and other expressions of not caring for "others." Addressing this involves, in David Bohm's words, paying attention to the whole situation so that we can "discover what is really an appropriate sort of action, relevant to the whole, for bringing the turbulent structure of vortices to an end."[5]

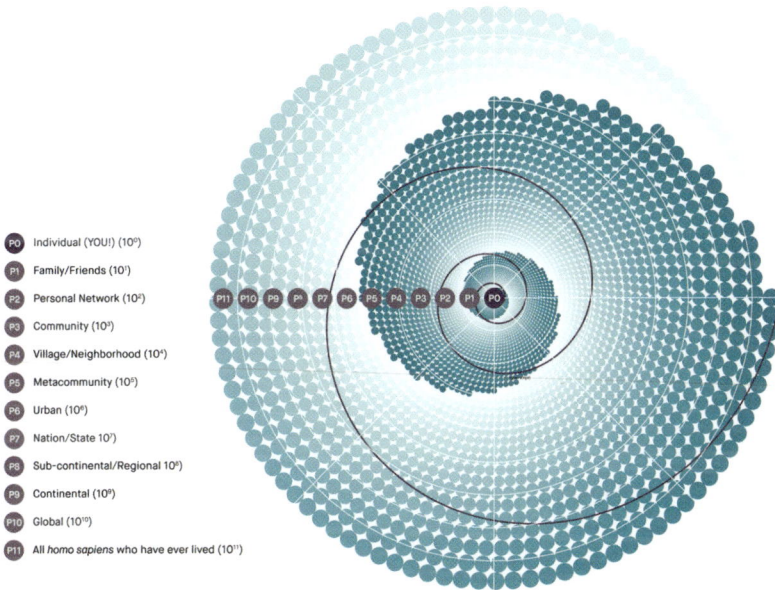

P0 Individual (YOU!) (10^0)

P1 Family/Friends (10^1)

P2 Personal Network (10^2)

P3 Community (10^3)

P4 Village/Neighborhood (10^4)

P5 Metacommunity (10^5)

P6 Urban (10^6)

P7 Nation/State 10^7)

P8 Sub-continental/Regional 10^8)

P9 Continental (10^9)

P10 Global (10^{10})

P11 All *homo sapiens* who have ever lived (10^{11})

Figure 4: P10 Fractal Framework (Source: McCaffrey 2020).

PRACTICE MATTERS

Still, a big question remains: What does quantum social change mean in practice?" Perhaps the answer to the question lies in the question itself. *Practice matters*. To "practice" is a verb that refers to the actual application or use of an idea, a belief, an intention, a theory, or a method. It also describes turning something into a procedural or habitual way of doing something. Social practice theory focuses on the relationship between individual actions and behaviors and socio-technical systems, including social norms, relations, and institutional contexts. Practices are viewed by sociologist Elizabeth Shove as "carried, sustained and transformed by cohorts of practitioners (those who do)."[6] In particular, social practice theory maintains that achieving social change requires a focus "on the dynamics of social practice, rather than 'behaviour change' narrowly defined," in that it is our continuous and repeated daily actions that anchor and generate that change.[7]

In everyday terms, practice is often seen as the pathway to a particular state – mastery in terms of a sport, talent, or state of being (as in, "the pathway to enlightenment"). For example, I notice that practicing the piano helps me to master a classical piece, like Beethoven's *Ode to Joy*. With no practice, I experience no mastery. Yet mastery involves more than a prerequisite number of practice hours. It is also, as I have learned, about *how* I show up. This too is a practice, but mastery can be experienced in the moment. As a practice and a process, quantum social change is about mattering from a balanced and coherent state of [I/we], here and now, and in the next instant, and the next, and so on. In short, we can think of "mattering" as a process and practice of holding a belief in an equitable and sustainable future in the present, and grounding actions in universal values that apply to all sentient and non-sentient beings. To collapses a wave of potential into an event or experience that resonates with the whole involves a continuous practice of conscious change, creative change, collaborative change, and courageous change.

Many spiritual traditions, embodied practices, and political movements emphasize the importance of bringing awareness and attention to [whole/parts], [I/we], and [both/and] relationships. This often involves drawing attention to relationships between humans and the environment, mind and body, and the individual and collective. For example, in terms of spiritual traditions, the

Haudenosaunee culture in North America recognizes that we *are* the environment, and so is society; they emphasize the importance of happiness, love, and achieving a state of balance between opposing forces to create a state of consensus and accord, or "one mind."[8] In terms of embodied practices, vipassana meditation emphasizes an awareness of the impermanent nature of being through a practice of noticing physical sensations without reacting to them. Practicing equanimity or inner peace while noticing desirable and undesirable sensations can contribute to an awareness of the deep links between mind and body, mind and matter, and "I" and others. Finally, as an example of a political movement, the Alternative U.K. is a platform that takes a fractal approach to politics by focusing on values such as courage, generosity, transparency, humility, humor, and empathy. Their approach aims to change the language and practice of politics:

> The values are not just there to be brought out on special occasions. The six core values must be constant indicators that are visible in our daily political work – in the way we think, speak and act. From debate, to political initiatives and to the way we campaign.[9]

As discussed in Chapter 3, believing that change is possible and that we have the potential to transform ourselves and our systems is a critical starting point for social change. Monica Sharma emphasizes this:

> Transformation is not simply transcending differences or finding common ground or new formations; it is new formations, new patterns and systems grounded in what we stand for, our actions emanating from embodying universal values such as compassion, equity, and dignity for everyone, everywhere.[10]

In an entangled quantum universe, values that apply to the whole will generate different impacts than those that apply only to parts, or as Karin Fierke writes, "From the perspective of entanglement and compassion, if I and thou are not entirely separate atomistic beings, then harm to another is ultimately harm to the self as well."[11]

THE ESSENCE OF CLIMATE CHANGE

This book has explored the potential for quantum social change, based on the idea that climate change is, at its very essence, about relationships. The point of this inquiry has not been to challenge the classical scientific understanding of the physical and social consequences of climate change. Instead, it has been to dig deeper into the ways that we understand ourselves and our capacity to respond to global crises by looking at what quantum social science has to say about relationships between mind, meaning, and matter. Quantum social change is about activating a quality of agency by responding from a space of integrity and oneness in every moment. The potential for quantum social change lies not in technologies, but in people, and it is expressed through love for ourselves, each other, nature, the planet, and future generations. From the perspective of quantum social change, [I/we] are the most powerful solutions to climate change.

Quantum social change is not just about choosing a different paradigm. It is about *being* a different paradigm. This takes courage in a world where we are socialized to conform and habituated to believe in the paradigm that we are separate. It involves holding paradigms flexibly and being open to new ideas. The power to transcend paradigms is considered by Donella Meadows to be the *highest* leverage point for systems change, and a source of radical empowerment:

> That is to keep oneself unattached in the arena of paradigms, to stay flexible, to realize that NO paradigm is "true," that every one, including the one that sweetly shapes your own worldview, is a tremendously limited understanding of an immense and amazing universe that is far beyond human comprehension. It is to "get" at a gut level the paradigm that there are paradigms, and to see that that itself is a paradigm, and to regard that whole realization as devastatingly funny.[12]

The diversity of interpretations described in this book are by no means complete or conclusive – they simply invite a deeper inquiry and wider engagement with the significance of quantum physics for the social world. It should at least be clear that it is worth questioning the assumption that quantum processes *do not* affect social life and social change. In fact, it might be more constructive to start with the question, *how could they not?* Tom Roberts argues that "as thinking beings in a processual universe we therefore have a certain ethical responsibility to foster experimentation with new modes of thought."[13] If we pursue these modes of thought, we may just discover that [I/we] share a remarkable capacity for social change.

Physicists and philosophers may end up debating interpretations and meanings of quantum physics for the next one hundred years. However, we are in the decade that matters, and the urgency of addressing global environmental change invites us to consider what it means for our reality, right here and now. Danah Zohar reminds us that "We actually live in a quantum world, and once we grasp this, nothing will ever be the same again."[14] Indeed, once we grasp the idea that we literally matter, and that we can individually and collectively contribute to a thriving world, we will see and be in a different world. Karen Barad puts this nicely:

> Different material intra-actions produce different materializations of the world, and hence there are specific stakes in how responsiveness is enacted. In an important sense, *it matters to the world how the world comes to matter.*[15]

What do quantum physics and quantum social science tell us about ourselves and our relationship to social change, such that we can generate an equitable and thriving world? Physicist John Archibald Wheeler once suggested that the central point of quantum theory was something that we should be able to express in one clear, simple sentence: "Until we see the quantum principle with this simplicity we can well believe that we do not know the first thing about the universe, about ourselves, and about our place in the universe."[16] Based on the inquiry presented in this short book, perhaps we can sum up a central point of quantum physics in one simple sentence: "You matter more than you think."

In short...

The scientific paradigm that has identified climate change as a problem is inadequate when it comes to motivating the individual and collective actions needed to respond to a world in crisis. If there is one thing that emerges from the inquiry presented in this book, it's the idea that we are underestimating our collective capacity for social change. Our current efforts to build connections, relate to diversity, and leverage change are classical struggles, and they are not succeeding in achieving the changes that are urgently needed. Quantum social change represents an alternative approach to scaling transformations. It is not an outcome, but a process and a practice that acknowledges that we are already connected and "co-related." When it comes to generating a thriving world for all, we each really do matter more than we think.

——— Reflections ———

- Do you agree with Danah Zohar that once we grasp that we live in a quantum world, nothing will ever be the same again? Why or why not? What would be different?

- What values do you stand for, and what visible changes do you want to see in the world?

- Recognizing that you are an important part of the solutions to climate change and other global crises, what can you do differently in the personal, political, and practical spheres of transformation?

EPILOGUE

QUESTIONS

QUESTIONS MATTER. They represent invitations to deeper inquiries that can help us to challenge some of the assumptions we hold about the nature of reality and our relationship to nature. Albert Einstein is often quoted as saying, "Don't listen to the person who has the answers; listen to the person who has the questions." The importance of asking questions, including about the nature of social world, cannot be understated. However, David Bohm also reminds us that "In scientific inquiries a crucial step is to ask the right question."[1] The right questions can open up new inquiries, leading to the discovery of new worlds.

We know that skepticism or doubt towards what is presented as knowledge or truth is critical to scientific advances, and it should be encouraged. In fact, a questioning of the individualistic, deterministic, and reductionist approaches to the social world inherited from Newtonian physics is what motivated the development of quantum social science in the first place.[2] This has further encouraged explorations into the relationships between individual change, collective change, and systems change.

Although skepticism can be healthy, directing it towards science in general can hinder responses to global issues such as climate change. For this reason, is important to make a distinction between open skepticism and closed skepticism. Open skepticism involves embracing uncertainty and maintaining a flexible mindset. As social scientist Helga Nowotny writes, "Uncertainty is the dynamic balance between what we know and do not yet know about the world and about ourselves."[3] In contrast, closed skepticism rejects any attempt at inquiry and is often based on a "right versus wrong" perspective that shuts down conversations. It does not recognize knowledge as an ongoing process of inquiry.

As an inquiry, the perspectives presented in this book should be approached with open skepticism. To engage with skeptics, I made a copy of the draft manuscript of the book available between June 2020 and March 2021. The book was downloaded by over 850 people, and I received feedback from more than 50 readers. A series of seven "You Matter" webinars were organized with experts and practitioners working with some of the themes discussed in the book. Feedback, dialogues, discussions, and my own reflections informed an extensive revision process. The questions and critiques were both constructive and challenging, and I directly address a few of them below.

An Open Inquiry

Question: I can see how quantum physics can be a source of metaphors and methods to inspire views and analyses social phenomena. However, I am not convinced of the merits of quantum social theory, and think that you place too much confidence in what physics can do and say about society – probably more than most physicists would.

Quantum social change is not just another effort to apply the logics of science and mathematics to the social world in this time of ecological crises. We should indeed be cautious about directly translating physical and mathematical theories to the social realm. However, quantum theory has already challenged some of the foundations of classical physics and positivist science, including the idea that subjects and objects are separate and distinct and that causality is always local. In fact, it may very well be showing us that physical science is also a social and human science.

Of course, there are many other ways of conceptualizing and expressing the role of entanglement and agency in processes of social change, without resorting to quantum physics or quantum metaphors.[4] Indeed, most of the ideas presented in this book have been said before by philosophers, social scientists, humanists, writers, and poets, and many of these ideas have been passed down through indigenous scholars and wisdom traditions from around the world.[5] Quantum social science adds to this by drawing attention to the potentiality that exists in every moment through the language of entanglement, superposition, complementarity, indeterminacy, and nonlocality. It offers a metaphorical bridge between the physical sciences and the social sciences and humanities, encouraging new ways to think about relationships between realism and idealism.[6] In *Globalization Development and Social Justice: A Propositional Political Approach*, human geographer Ann El Khoury suggests that quantum social science contributes to a reality that is "open and unfinished rather than closed and fixed, and has both an ideational and a material basis."[7] The power of the ideas embedded in quantum social theory draws attention to agential realism, participatory realism, and other approaches that recognize the significant role of individuals in processes of transformative change.

Question: Your argument about changing thinking and changing practice is rather ahistorical and ageographical, and so what about the real, variable constraints many people face in thinking and acting anew in places such as Russia, Brazil or Myanmar?

This is an important question, not the least because quantum concepts such as indeterminism, nonlocality, entanglement, and the relationship between subjects and objects have profound implications for understandings of geographical and historical phenomena.[8] Geographical and historical contexts influence how society is currently organized, who holds power, who or what is valued, and so on. A deterministic approach to our current social world would say, "this is just the way it is," implying that the situation is unlikely to change, except perhaps through revolutionary struggles. As Paolo Freire notes, "the oppressors develop a series of methods precluding any presentation of the world as a problem and showing it rather as a fixed entity, as something given – something to which people, as mere spectators, must adapt."[9] However, in Russia, Brazil, Myanmar, and elsewhere, many people have been persistently challenging hierarchical and deterministic worldviews presented by those in power, including those that are packaged in populist rhetoric. As Arturo Escobar describes in *Pluriversal Politics: The Real and the Possible*, individuals, groups, and communities around the world are challenging the politics of separation and fragmentation by practicing a politics of radical relationality.[10] When people in positions of power are confronted by an emancipatory politics, they often use all means possible to subjugate those who challenge them. This is often done through violent means, where the goal is to keep people in line, forcing them to passively adapt to the status quo.

A quantum approach to social change recognizes that local context is always entangled with a global context through language, meaning-making, and the perpetuation of myths that legitimate the sale of weapons, the training of military forces, the use of violence, the exploitation of labor and resources, the dehumanization of people, the extinction of species, and the pollution and destruction of the atmosphere, oceans, forests, water bodies, and soils. This draws attention to relationships, and to practicing "right relations" as a means of both solidarity and emancipation. In many historical and geographical contexts, the social wave function has collapsed into a reality that reflects nothing close to right relations. Yet, as Matthew Schneider-Mayerson and Brent Ryan Bellamy point out in *An Ecotopian Lexicon*, "our future is established not only through dramatic historical events but also through gradual accretion: moment by moment, act by act, word by word."[11] Today we are seeing that more and more people, especially within younger generations, are committed both to alternative ways of relating to each other and the planet, and to practicing the expression of [I/we] values from moment to moment. Through horizontal emergence that transcends scales, individuals can generate powerful fractals of change.

Question: *Why not focus social change on key people within powerful institutions instead of individuals in general?*

One of the key points that emerges from a quantum approach to social change is that every person has the capacity to shift systems and cultures. This is not the individualism of the classical world, but the [I/we] and [both/and] perspective of the quantum world. Key people within powerful institutions are more likely to retain habitual patterns that protect their interests, collapsing a wave of potential into a reality that maintains rather than transforms the status quo. Individuals within communities, organizations, or institutions who support transformative change are often reprimanded, excluded, expelled, tortured, or killed as a means of ensuring that people "stay in line."

Although the progressive end of a "potentiality distribution" aims to shift cultures and systems in the direction of a more reso-nant social consciousness, this tends to be approached from an "us versus them" perspective, with "them" referring to those who accept the current situation as a given, or as a privilege to be protected. While such polarization can be a mobilizing force, this "othering" tends to focus on individuals, rather than on the intra-actions that are perpetuating inequitable and unsustainable cultures and systems. We know that equity and unity are deep-seated values that drive collective change at every scale, includ-ing via social movements. Unleashing the power of individuals as agents of change does not exclude changes from emanating within powerful institutions or among those who have a large sphere of influence. Rather, it recognizes that this alone is not enough, and that *all* people can be fractals of change that generate new patterns and relationships.

Question: Is the lack of sustained examples proof positive that your arguments are idealistic or utopian?

The lack of sustained examples of societies that are organized to benefit both people and the planet may be a testimony to the failure of the classical view of the world, more than proof of the limits of a quantum worldview. There is a long history of indigenous cultures who have lived in harmony with the environment, and the march of history has been an arduous journey toward the embrace of ever-larger circles of care. In describing the emergence of an empathic civilization, Jeremy Rifkin ironically notes that "climate change is forcing us, as never before, to recognize our shared humanity and our common plight, in an essential way, rather than a superficial way."[12] However, until idealism and utopianism are recognized as potentials rather than delusions, most people will continue to "bet" that such realities are improbable. In *Envisioning Real Utopias*, sociologist Eric Olin Wright emphasizes that we need to create "real utopias" based on ideals that are grounded in the real potentials of humanity, yet which also help us navigate a world of imperfect conditions for social change.[13] We know that millions of actions are taken every day towards this goal, creating an invisible geography that is seldom included within the conventional bounds of knowledge, i.e., parallel worlds that are very real, yet not part of the classical script that most people are taught to live by.[14] However, this geography does not have to remain invisible. As Patrick Baert writes, stepping outside of our taken-for-granted world "empowers people to develop their imaginative abilities in that they become able to conceptualize what is not present."[15]

Question: Your critique of the classical, materialist depiction of humans is general, and does not acknowledge how this is mediated through textbooks, teaching practices, and culture at large. What purpose does such a straw man argument serve?

The point of this inquiry is not to create a straw man, but to acknowledge that the "straws" of the classical paradigm have indeed been woven tightly into textbooks, teaching practices, and culture at large, to the extent that it is threatening the sustainability of the planet. Our assumptions about relationships to ourselves, each other, the natural world, the climate system, and the future are influenced by a view of the world that sees subjects and objects as distinct and separate. In *Nature's Economy: The History of Ecological Ideas*, historian Donald Worster explores how ideas about the natural world are rooted in culture. Starting with the Age of Reason, he traces the evolution of ideas about nature toward today's understanding that "if the observer always affects the observed, changing it from moment to moment, from glance to glance, then the observed also changes the observer."[16] It is important to be aware of how different understandings and ideas influence teaching practices, and what quantum social science implies for education. Though there are many cultural interpretations of science and society, it is always worth considering how basic understandings of reality shape our approach to social change.

Question: Isn't it dangerous to dismiss classical physics in favor of a quantum approach to social science? Is an inquiry into quantum social change really necessary and worthwhile?

Living within a classical paradigm that perpetuates disconnected relationships with ourselves, each other, and the environment, it is easy to stand aside and watch our planet warm, count species as they disappear, and act as if there is nothing we can really do about it. The classical paradigm fails to ignite the sparks of agency needed to manifest the quality and depth of changes called for in times of crisis. When we accept classical physics as the underlying basis for our social reality, we limit our perceived capacity to collectively influence the world in a positive way. If we cannot influence outcomes at scale, can we truly take responsibility for the future?

The problem is not with classical physics itself, for there is no doubt that locality, causality, and determinism are relevant to most physical aspects of our lived experience. For example, as temperatures increase, ice melts and water expands; the melting of ice sheets and warming of oceans will raise sea levels, posing risks to coastal communities and ecosystems. Yet there are significant differences between an average global warming of 1.5°C and a 2°C, 3°C or 4°C warming, just as there are enormous differences between 50 centimeters of sea level rise and one, two, three or more meters. The outcomes and impacts will be influenced by the types and qualities of individual and collective responses. Classical physics tells us that mitigation is critical, adaptation is necessary, and transformation is essential.

If there is one thing that climate change tells us, it is that humans can influence global systems. Individually and collectively, our decisions and actions are contributing to increased concentrations of greenhouse gases in the atmosphere, changes in land cover, depletion of soils, degradation of ecosystems, and many other environmental changes. We are shifting patterns and transforming systems at a global scale. Recognizing this, we are confronted with the challenge of intentionally transforming ourselves and our economic, social, and political systems so that all life can thrive. To realize this, we have to develop the capacity to both "see change" and "be change" in a different way. Quantum social science offers a relational lens on social change and recognizes that small changes can make a big difference, particularly when we embrace concepts such as non-locality, entanglement, complementarity, superposition, and potentiality.

In describing the principle of pragmatism, psychologist William James suggests that an inquiry is worthwhile if it makes a difference in the world.[17] If theories can be seen as instruments rather than answers, then quantum social theory offers an opportunity to both explore and experiment with how an alternative paradigm can influence social change. In fact, we can go further and actually test it to see whether it makes a difference in the world. In "the decade that matters," now is a good time to explore whether we can make a difference at scale and to experiment with quantum social change. As Hans Christian von Baeyer reminds us, in quantum mechanics, "*unperformed experiments have no outcomes.*"[18]

Question: Paradigms take a long time to change, and we do not have time. Given the urgency, shouldn't we just focus on political solutions and social movements?

The urgency of this moment has been a key motivation for this inquiry into what quantum social science can offer to understandings of social change. When it comes to responses, this decade matters, and the importance of generating a quantum leap to sustainability cannot be understated. Yet urgency may not be the most effective driver of social change. Indeed, climate scientists Amanda Lynch and Siri Veland warn us that "any discourse of urgency has the potential to both expand and limit the perceived range of alternative responses."[19] To realize the potential that exists here and now, Monica Sharma emphasizes the importance of holding a space of urgency, without holding the stress of urgency.[20] Political solutions and social movements are important, but the classical approach to social change is likely to perpetuate what philosopher John Foster refers to as "the politics of never getting there."[21]

A quantum approach to social change recognizes nonlocality and entanglement, and is based on the idea that [I/we] are part of a much larger social movement. In *Blessed Unrest*, environmentalist

Paul Hawken describes how more than a million organizations working towards sustainability and social justice together make up the largest social movement in history.[22] The movement is already underway, and it can be amplified through the quality of our individual agency, collective agency, and political agency. If we take the millions of examples of people caring for each other, for equity and social justice, for biodiversity, and for a climate where life on the planet can thrive, we see that quantum fractals are everywhere. Together, they form what Ann El Khoury refers to as different ways of relating to social change: an invisible "infraglobalization" that can contribute to bringing alternative spaces, practices, and possibilities for the future into the present.[23] The point here is that it is possible to generate an equitable and thriving world, one quantum fractal at a time. Being open to the potential for quantum social change invites an ethical stance, and we are encouraged to take responsibility for the role that we play in shaping our universe.[24]

Question: I would like to believe all of this, but I simply don't see much potential for transformative change right now, given the situation in the world today.

The potential for change exists in the world today. As author and activist Naomi Klein emphasizes in her book *This Changes Everything,* climate change is the problem that will transform the world one way or another. However, she also points out that it may be our best chance to transform the world for the better.[25] The COVID-19 pandemic and its secondary effects have made this call for a better world even more actual and urgent. Which paradigm of social change will we choose to transform the world? Which stories will we tell ourselves about our future? What types of relationships will we prioritize?

There is a growing recognition of entanglement, complementarity, indeterminacy, and potentiality, led by those who sense or have experienced that we live in an entangled universe, and are intra-acting on behalf of [I/we] and [whole/parts]. Although the responses to climate change and other complex global problems have largely focused on technical and behavior approaches to social change, many people are ready to "matter" in deeper and more meaningful ways. As discussed in this book, quantum physics may have something to offer us, including metaphors, methods, and meanings that emphasize our potential to matter. Physicist and feminist Shohini Ghose argues that "nature has shown us that you can take the smallest amounts—trace elements of matter—and they can actually make huge impacts and change the planet. So can we be those trace elements and make tiny actions that collectively change the world?"[26] In the decade that matters, how we perceive and activate our potential to scale change will be decisive. A quantum perspective offers new ways of thinking about this, opening our minds to the possibility and potential that exists here and now. The future that we experience will depend on whether and how we actualize this potential for social change.

A CONTINUOUS INQUIRY

Where do we go from here? How can we be or become fractals of quantum social change? These are important questions, and although the answers will depend on each one of us and our particular context, it is important to remember that the questions themselves actually matter too. There is a need for continuous inquiry, reflection, and action. The goal of this book has been to explore emerging theories and propositions about the nature of the social world based on new insights from quantum social science. This approach to social science draws attention to how we relate to change itself, and specifically to the role that each of us plays in transformative change. We have considered the possibility that social change can be approached through the metaphors and meanings of quantum physics, which together contribute to new ways of thinking about the relationship between individual change, collective change, and systems change. Above all, we have explored an alternative way of scaling social change, based on the recognition that [I/we] can generate new patterns and relationships that matter. In *Life on the Edge: The Coming of Age of Quantum Biology*, Jim Al-Khalili and Johnjoe McFadden point out that "Quantum mechanics is normal. It is the world it describes that is weird."[27] The task ahead is to make quantum social change normal. After all, there is nothing weird about generating a world where all life can thrive.

CONNECTION

Let us be quantum,
entangled across spacetime,
hearts and minds as one.

—Shohini Ghose

ACKNOWLEDGMENTS

This book has been a passion project – one that I have felt compelled to work on at the oddest hours and in the strangest places, often without knowing exactly why. Okay, it's been more like an obsession, and it's demanded patience from a lot of people! I feel so much gratitude to everyone those whose enthusiasm and support have kept me going over the past years, and particularly to those who provided feedback on the draft version that I released in 2020. I'd also like to thank everyone who participated in the You Matter webinars series – they were bright spots for me during the COVID-19 lock-down. And I would never have gotten through a long winter and the substantial revision process without playing and replaying the song "Wildflower Meadow" by The Eternal Page. This is my son's band, and the name truly resonates with this writing project.

My interest in quantum social science was sparked a decade ago, when I started thinking about the role of paradigms in social change and systems change. I came across Alexander Wendt's work on quantum social theory and I'm grateful to him for being generous with his time and his work. I would also like to thank my long-time collaborator and friend, Robin Leichenko, for joining me on a "quantum road trip" to Ohio State University to talk to Alex about his ideas and consider how they relate to climate change. I'm grateful to Ann El Khoury for the opportunity to exchange ideas around our mutual interest in quantum social science and its implications for justice. I'd also like to thank James Der Derian for inviting me to the 2017 "Q Symposium" in Sydney, which is part of *Project Q: Peace and Security in a Quantum Age*. It was great to learn that so many fascinating people are exploring the social and political implications of quantum physics.

It's been an honor and a privilege to work with the Adaptation-CONNECTS research group, my cCHANGE colleagues, and so many others who have been interested in these ideas, providing suggestions on the text, as well as encouragement, practical assistance, and support. This includes Leonie Goodwin, Linda Sygna, Morgan Scoville-Simonds, Gail Hochachka, Irmelin Gram-Hanssen, Milda Rosenberg, Emma Arnold, Julia Bentz, Ann Kristin Schorre, Teresia Aarskog, Dina Hestad, Danielle Huffaker, Nora Engeseth, Annabel Mempel, Heidi Bragerhaug, Randi Larsen, Ingvild Nilsson, Nicole Schafenacker, Cyrille Rigolot, Leonardo Orlando, Ton Baggerman, Mark McCaffrey, David Tabara, Noel Castree, Steven Hartman, Erik Knain, Mark Edwards, Angela Frydenger, Susan Jackman, Camilla Sundsbak, Asun St. Clair, Diana Liverman and so many others. I am also grateful to Anne Caspari, Elke Fein, Bettina Geiken, Indra Adnan, Sandra Waddock, Chris Riedy, Simon Divecha, Jordan Rosenfeld, Thomas Hübl, Terry Patten, Michael Stone, Petra Kuenkel, Ute Thiermann and many others for stimulating ideas, conversations, and inspiration. Finally, I appreciate the many students whose curiosity about "quantum social change" has encouraged my inquiry, as well as the Research Council of Norway and the University of Oslo for funding the AdaptationCONNECTS project.

A big thanks to Christina Bethell for writing a wonderful foreword to the book – her work on the importance of mattering has been so inspiring. Also, many thanks to Nada Qamber of Qamber Designs for her excellent work on the book's interior and formatting – and her patience with my endless changes. My deepest gratitude goes to Anne Pender for her incredible support, vision, and editing skills in the writing process. Her wide range of knowledge, sharp eye, shared interest in this topic, and generosity has made such a difference. I'm also grateful to Janine Stenehjem, who challenged me back in 2015 to write a "short and accessible" book about quantum social change. It only took me six years! It's not as short as I had hoped, but this inquiry is only a starting point for what is truly an "eternal page."

Tone Bjordam has made a tremendous difference to this project by sharing her amazing artistic talents. It has been wonderful to collaborate with her and watch her creative ideas come to life as artwork. I also appreciate valuable insights, mentoring, and feedback from Monica Sharma. Working with her over the past decade has helped me to see what quantum social change means in practice, and why agency that is based on equity, dignity, and compassion matters. I would also like to thank Barbara O'Brien, Tom and Pati O'Brien, Kenneth O'Brien and Karen Grosskreutz for endless encouragement and support. Finally, I'm grateful to Kristian, Jens Erik, Espen, and Annika Stokke for their love and inspiration – you all matter so much to me!

ENDNOTES

FOREWORD

1. Trauma and toxic stress ACEs can dramatically impact early and lifelong physical, mental, and behavioral health and predict early death. They also underlie the sense of disconnection, inequities and social unrest pervading our world today. Without awareness and action to heal the trauma we carry, ACEs are perpetuated across generations until eventually the syndemic of collective trauma, mental and physical illness, and loss of mattering and meaning can inhibit or impede the changes required to heal ourselves, our environment, and our climate (see Bethell et al. 2017; also Hübl 2020).
2. See Bethell et al. 2019.

INTRODUCTION

1. For example, see Haven and Khrennikov, *Quantum Social Science*; Wendt, *Quantum Mind and Social Science: Unifying Physical and Social Ontology;* and O'Brien, "Climate Change and Social Transformations: Is It Time for a Quantum Leap?"
2. Steffen et al., "Trajectories of the Earth System in the Anthropocene."
3. See IPCC, *Global Warming of 1.5°C.*
4. See IPBES *Global Assessment.*
5. Figueres and Rivett-Carnac, *The Future We Choose: Surviving the Climate Crisis.*
6. The Doomsday Clock is a metaphor established by the Bulletin of the Atomic Scientists to represent how close the world is to a global catastrophe. https://thebulletin.org/doomsday-clock/current-time/
7. Burwell, *Quantum Language and the Migration of Scientific Concepts.*
8. Wendt, 33.

CHAPTER 1: THIS DECADE

1. Horgan, *Mind-Body Problems: Science, Subjectivity & Who We Really Are* (Introduction).
2. These perspectives have been raised by numerous researchers, yet have not been taken seriously within the Earth System paradigm; see Waddock and Kuenkel, What Gives Life to Large System Change?" and O'Brien, "Reflecting on the Anthropocene: The Call for Deeper Transformations."
3. Brown, *Plan B 4.0: Mobilizing to Save Civilization;* Rockström et al. "A Roadmap for Rapid Decarbonization."
4. For example, see Bradford, *Save the World: There is No Planet B. Things You Can Do Right Now to Save our Planet;* and https://davidsuzuki.org/what-you-can-do/top-10-ways-can-stop-climate-change/
5. Hawken, *Drawdown: The Most Comprehensive Plan Ever Proposed to Reverse Global Warming;* Frischmann et al., "Drawdown's 'System of Solutions' Helps to Achieve the SDGs."

6. Jackson et al. *Global Energy Growth Is Outpacing Decarbonization.*
7. See Griffin, *CDP Carbon Majors Report 2017*; Rainforest Action Network, Banking on Climate Change 2020; and https://ec.europa.eu/clima/policies/transport/aviation_en
8. The Shift Project, *Lean ICT: Towards Digital Sobriety.*
9. Leichenko and O'Brien, *Climate and Society: Transforming the Future*; Roberts and Parks, *A Climate of Injustice: Global Inequality, North-South Politics and Climate Policy.*
10. Leichenko and O'Brien, *Environmental Change and Globalization: Double Exposures.*
11. Leichenko and O'Brien, *Climate and Society.*
12. Wendt, *Quantum Mind and Social Science.*
13. Escobar, *Designs for the Pluriverse: Radical Interdependence, Autonomy, and the Making of Worlds*, 170.
14. Joronen and Häkli, "Politicizing Ontology."
15. See Feola, "Societal Transformation in Response to Global Environmental Change: A Review of Emerging Concepts."
16. Sharma, *From Personal to Planetary Transformation*, 34.
17. See Hord and Lee (eds.), *I Am Because We Are: Readings in Africana Philosophy*; Allison, "The Spiritual Significance of Glaciers in an Age of Climate Change"; and Berkes; *Sacred Ecology.*
18. Deloria, *The Metaphysics of Modern Existence*; Capra, *The Tao of Physics.*
19. Coole and Frost, *New Materialisms: Ontology, Agency, and Politics*, 29.
20. Bennett, *Vibrant Matter: A Political Ecology of Things*, 25.
21. Rosiek et al., "The New Materialisms and Indigenous Theories of Non-Human Agency: Making the Case for Respectful Anti-Colonial Engagement," 332.
22. Weber, *Enlivenment: Towards a Fundamental Shift in the Concepts of Nature, Culture and Politics.*
23. Ibid.
24. Fuchs, "On Participatory Realism," 3.

CHAPTER 2: PARADIGMS

1. In "Leverage Points: Places to Intervene in a System," Meadows emphasizes that "Paradigms are the sources of systems. From them, from shared social agreements about the nature of reality, come system goals and information flows, feedbacks, stocks, flows and everything else about systems." (p.18).
2. Kuhn, *The Structure of Scientific Revolutions*, 153.
3. For example, see Crease and Goldhaber, *The Quantum Moment: How Planck, Bohr, Einstein, and Heisenberg Taught Us to Love Uncertainty* and Gilder, *The Age of Entanglement: When Quantum Physics Was Reborn.*
4. Stapp, *Mindful Universe: Quantum Mechanics and the Participating Observer*, 7.
5. Kleppner and Jackiw, "One Hundred Years of Quantum Physics," 893.
6. For brief explanations of these terms, see https://en.wikipedia.org/wiki/Glossary_of_quantum_philosophy
7. Kaiser, *How the Hippies Saved Physics: Science, Counterculture, and the*

Quantum Revival.

8. Proietti et al. "Experimental Test of Local Observer Independence."

9. The BIG Bell Test Collaboration, "Challenging Local Realism with Human Choices."

10. See Chen, "Even Huge Molecules Follow the Quantum World's Bizarre Rules;" Kotler et al., "Direct Observation of Deterministic Macroscopic Entanglement;" and Fein et al., "Quantum Superposition of Molecules Beyond 25 kDa."

11. Rovelli, *Reality Is Not What It Seems: The Journey to Quantum Gravity.*

12. Barad, *Meeting the Universe Halfway: Quantum Physics and the Entanglement of Matter and Meaning,* 67 (italics in original).

13. Ibid.

14. Nagel, *The View From Nowhere.*

15. Carroll, *Something Deeply Hidden: Quantum Worlds and the Emergence of Spacetime.*

16. Interestingly, the Many-Worlds Interpretation does not seem to imply that our conscious choices and decisions actually matter. In *Something Deeply Hidden*, physicist Sean Carroll argues that branching is the result of a microscopic process amplified to macroscopic scales, whereas a decision is a macroscopic phenomenon, not something made by microscopic electrons or atoms in the brain. Of course, this is assuming that neural processes are themselves purely classical phenomena.

17. Fuchs, "On Participatory Realism."

18. Ibid., 1.

19. Mermin, "Physics: QBism Puts the Scientist Back Into Science."

20. Stapp, 36.

21. Ibid., 7.

22. Ibid., 4.

23. Wendt, 4.

24. Rovelli, *Seven Brief Lessons on Physics,* 76.

25. Hickman, "James Lovelock: Humans are too stupid to prevent climate change."

26. Franzen, "What if we stopped pretending?"

27. Watts and Stenner, *Doing Q Methodological Research: Theory, Method and Interpretation.*

28. Wendt.

29. Kaiser.

30. Kaiser; Capra.

31. Der Derian and Wendt, " 'Quantizing International Relations': The Case for Quantum Approaches to International Theory and Security Practice."

32. For example, see Alonso-Sanz, *Quantum Game Simulation*; Busemeyer and Bruza, *Quantum Models of Cognition and Decision*; Danilov and Lambert-Mogilliansky, "Preparing a (Quantum) Belief System;" and Orrell, *Quantum Economics: The New Science of Money.*

33. Busemeyer and Bruza, *xii.*

34. Haven and Khrennikov, 54.

35. Wendt.
36. Barad.
37. Al-Khalili and McFadden, *Life on the Edge: The Coming of Age of Quantum Biology*, 58.

CHAPTER 3: BELIEFS

1. Rovelli, *Reality Is Not What It Seems*, 231.
2. Nilsson, *Understanding Beliefs*.
3. Milkoreit, *Mindmade Politics: The Cognitive Roots of International Climate Governance*, 239.
4. Nilsson; Milkoreit.
5. Stoknes, *What We Think About When We Try Not To Think About Global Warming*.
6. Bendell, "Deep Adaptation: A Map for Navigating Climate Tragedy," 19.
7. von Baeyer, *QBism: The Future of Quantum Physics*.
8. Keller, "Bayesian Decision Theory and Climate Change."
9. Hobbs, "Bayesian Methods for Analysing Climate Change and Water Resource Uncertainties."
10. Meadows et al., *The Limits to Growth: A Report for the Club of Rome's Project on the Predicament of Mankind*, 29.
11. Ibid.
12. Schlitz et al., "Worldview Transformation and the Development of Social Consciousness,"
13. Xie et al., "Social Consensus through the Influence of Committed Minorities."
14. Ball, *Critical Mass: How One Thing Leads to Another*.
15. Ibid., 95.
16. Barad, 361.
17. Hobbs.
18. Tebaldi et al., "Quantifying Uncertainty in Projections of Regional Climate Change: A Bayesian Approach to the Analysis of Multimodel Ensembles," 1536.
19. Wendt, 47.
20. Ibid., 48.
21. Barad, 261.
22. Ibid., 35.
23. von Baeyer, 127.
24. Fuchs and Stacey, "QBism: Quantum Theory as a Hero's Handbook," 1.
25. Milman, "Planet has just 5% chance of reaching Paris climate goal, study says."
26. Raftery et al., "Less than 2°C Warming by 2100 Unlikely."
27. Ibid., 4.
28. Otto et al., "Human Agency in the Anthropocene."
29. Johnson, *The Body in the Mind: The Bodily Basis of Meaning, Imagination, and Reason*.
30. van der Kolk, *The Body Keeps the Score: Brain, Mind, and Body in the Healing of Trauma*.
31. von Baeyer.

CHAPTER 4: RELATIONSHIPS

1. Steffen et al., "Planetary Boundaries: Guiding Human Development on a Changing Planet."
2. Leichenko and O'Brien, *Environmental Change and Globalization.*
3. Wendt.
4. Ball, *Critical Mass*, 568.
5. Gunderson and Holling, *Panarchy: Understanding Transformations in Human and Natural Systems.*
6. Alaimo, *Bodily Natures: Science, Environment, and the Material Self*; Oliver, *The Self-Delusion: The Surprising Science of Our Connection to Each Other and the Natural World.*
7. O'Neil, *Weapons of Math Destruction: How Big Data Increases Inequality and Threatens Democracy*, 38.
8. Barad, 33.
9. Ibid.
10. Bohm, *Wholeness and the Implicate Order*, 52.
11. Wendt, 210.
12. Ibid., 217.
13. Ibid., 258 (italics in original).
14. Ibid., 264.
15. Ibid., 265.
16. Ibid., 264.
17. Ibid., 265.
18. Barad, 393.
19. Burwell, 256.
20. Wilber, *Sex, Ecology, Spirituality: The Spirit of Evolution*, 43.
21. Ibid., 117.
22. Ibid, 43.
23. For an elaboration on the distinction between dualism and duality, see Jackson, "Dualism, Duality and the Complexity of Economic Institutions" and Straw and Ison, "Duality, dualism, duelling and Brexit."
24. Edward Sapir, 564.
25. Siegel, *IntraConnected: MWe (Me + We) as the Integration of Belonging and Identity.*
26. Yunkaporta, *Sand Talk: How Indigenous Thinking Can Save the World*, 85.
27. Gram-Hanssen et al., "Decolonizing Transformations Through 'Right Relations'."
28. Barad, 391.
29. Esposito, "In Lak'ech—a la K'in."
30. Ibid., 132.

CHAPTER 5: METAPHORS

1. Lakoff and Johnson, *Metaphors We Live By*, 7.
2. Ibid., 146.
3. Shaw and Nerlich, "Metaphor as a Mechanism of Global Climate Change Governance: A Study of International Policies, 1992–2012."

4. Bohm and Peat, *Science, Order and Creativity*, 35.
5. Taylor and Dewsbury, "On the Problem and Promise of Metaphor Use in Science and Science Communication," 1.
6. Lakoff and Johnson, *Metaphors We Live By*, 158.
7. Bohm and Peat, 50.
8. Lakoff and Johnson, *Philosophy in the Flesh: The Embodied Mind and Its Challenge to Western Thought*.
9. Barad.
10. Ibid., 182.
11. Ibid., 269.
12. Wendt, 258.
13. Ibid.
14. Burwell.
15. Alexander, *The Jazz of Physics: The Secret Link Between Music and the Structure of the Universe*.
16. Ibid., 113.
17. Ibid., 216.
18. Ibid., 40.
19. Ibid., 215.
20. Ibid., 175.
21. Ibid., 165.

CHAPTER 6: ENTANGLEMENT

1. Barad, 270 (italics in original).
2. Barad, 392.
3. Konvalinka et al., "Synchronized Arousal between Performers and Related Spectators in a Fire-walking Ritual."
4. Ferrari and Coudé, "Mirror Neurons, Embodied Emotions, and Empathy."
5. Lieberman and Eisenberger, "Pains and Pleasures of Social Life," 891.
6. Lieberman, *Social: Why Our Brains Are Wired to Connect*, 96.
7. Gisin, *Quantum Chance: Nonlocality, Teleportation and Other Quantum Marvels*, 108.
8. Ibid., 78. It should be noted that even within this window of twelve years, Einstein's theory challenged our very understanding of locality; showing that space does not actually support *any* localized structure (see Musser, "How Einstein Revealed the Universe's Strange 'Nonlocality'").
9. Gisin.
10. Barad, 393.
11. Bohm, 59.
12. Bradley, *The Cambridge World History of Slavery, AD 1420 - AD 1804*, 164.
13. See Lynch, "The Role of the Yorta Yorta People in Clarifying the Common Interest in Sustainable Management of the Murray-Darling Basin, Australia"; Asch and Macklem, "Aboriginal Rights and Canadian Sovereignty: An Essay on R. v. Sparrow."
14. Ball, *Beyond Weird: Why Everything You thought You Knew about Quantum*

Physics is Different, 162.

15. Wendt.
16. Kahane, *Collaborating with the Enemy: How to Work with People You Don't Agree with or Like or Trust.*
17. Wendt, 33.
18. Ibid., 220.
19. Brown, "A Primer on Q Methodology."
20. Watts and Stenner, *Doing Q Methodological Research.*
21. Watts and Stenner, "Q Methodology, Quantum Theory, and Psychology," 165 (italics in original).
22. Ibid., 169 (italics in original).
23. Ibid., 170 (italics in original).
24. Ibid.
25. Ibid., 168.
26. Cameron, "New Geographies of Story and Storytelling," 580.
27. Phillips, "Storytelling in Earth sciences: The Eight Basic Plots."
28. Foust and Murphy, "Revealing and Reframing Apocalyptic Tragedy in Global Warming Discourse."
29. Berry, "The New Story," 12.
30. Eisenstein, *Climate: A New Story*, 260.
31. O'Brien et al., *Our Entangled Future: Stories to Empower Quantum Social Change.*
32. Riedy, "The Witnesses," 15.
33. Johnson, *The Body in the Mind*, 209 (emphasis in original).
34. Zanotti, "De-colonizing the Political Ontology of Kantian Ethics: A Quantum Perspective," 11.

CHAPTER 7: CONSCIOUSNESS

1. Freire, *Pedagogy of the Oppressed*, 79.
2. Wilber, *Sex, Ecology, Spirituality*, 117.
3. Nagel, *Mind and Cosmos: Why the Materialist Neo-Darwinian Conception of Nature Is Almost Certainly False*, 53.
4. Zohar and Marshall, *The Quantum Society: Mind, Physics, and a New Social Vision*, 28.
5. Teilhard de Chardin, *The Phenomenon of Man.*
6. Ibid., 181.
7. Samson and Pitt, *The Biosphere and Noosphere Reader: Global Environment, Society and Change*, 2.
8. Wilber, *Sex, Ecology, Spirituality*, 100.
9. Haff, "Humans and Technology in the Anthropocene: Six Rules."
10. Rosol et al. "Introduction: In the Machine Room of the Anthropocene."
11. See Dennett, *Freedom Evolves* or Torey, *The Conscious Mind.*
12. Wilber, *Sex, Ecology, Spirituality.*
13. Wilber, *Eye to Eye: The Quest for the New Paradigm.*
14. Freire, *Pedagogy of the Oppressed*, 130.
15. Freire, *Pedagogy of Hope.*

16. Wilber, *Sex, Ecology, Spirituality.*
17. Esbjörn-Hargens, "An Ontology of Climate Change: Integral Pluralism and the Enactment of Multiple Objects."
18. Hochachka, "On Matryoshkas and Meaning-Making: Understanding the Plasticity of Climate Change," 10.
19. Ibid., 8.
20. Siegel, *The Mindful Brain: Reflection and Attunement in the Cultivation of Well-Being.*
21. Kegan and Lahey, *Immunity to Change: How to Overcome It and Unlock the Potential in Yourself and Your Organization.*
22. Kegan and Lahey, 21.
23. Schlitz et al.
24. Ibid.
25. Wendt, 128.
26. Freire, *Pedagogy of the Oppressed.*
27. Wendt.
28. See Koch, *The Feeling of Life Itself: Why Consciousness is Widespread but Can't be Computed;* and Tononi and Koch, "Consciousness: Here, There and Everywhere?"
29. Mathews, *For Love of Matter: A Contemporary Panpsychism,* 4-6.
30. Wendt.
31. Ibid., 125.
32. Wendt.
33. See Fierke, "Consciousness at the Interface: Wendt, Eastern Wisdom and the Ethics of Intra-Action" and Burgess, "Science Blurring Its Edges into Spirit: The Quantum Path to Ātma."
34. Wilber, *Quantum Questions: Mystical Writings of the World's Great Physicists.*
35. Kuhn, *The Structure of Scientific Revolutions.*

CHAPTER 8: AGENCY

1. See Wendt; Barad.
2. Zanotti, *Ontological Entanglements, Agency and Ethics in International Relations: Exploring the Crossroads,* 75.
3. Wilkerson, *Caste: The Origins of Our Discontents,* 184.
4. Baert, *Philosophy of the Social Sciences: Towards Pragmatism.*
5. See Bhaskar et al., *Metatheory for the Twenty-First Century: Critical Realism and Integral Theory in Dialogue.*
6. Doyle et al., "Amplified Melt and Flow of the Greenland Ice Sheet Driven by Late-summer Cyclonic Rainfall."
7. Mooney, "This is Climate Skeptics' Latest Argument about Melting Polar Ice – And Why it's Wrong."
8. Lorenzoni and Hulme, "Believing Is Seeing: Laypeople's Views of Future Socio-Economic and Climate Change in England and in Italy."
9. See Stoknes; Marshall, *Don't Even Think About It: Why Our Brains Are Wired to Ignore Climate Change.*

10. Hamilton, *Earthmasters: The Dawn of the Age of Climate Engineering*.
11. O'Neil, 197.
12. Boyd et al., "The 'New' Carbon Economy: What's New?"
13. Sharma, *Radical Transformational Leadership: Strategic Action for Change Agents*, 307.
14. Freire, *Pedagogy of the Oppressed*.
15. Monica Sharma, personal communication, February 13, 2020.
16. Wendt, 270.
17. Ibid.
18. Stapp, 5.
19. Scoville-Simonds, "Climate, the Earth, and God – Entangled Narratives of Cultural and Climatic Change in the Peruvian Andes."
20. Baggerman, *It's About Us: Meaning, Emotions and Mental Health in Post-Newtonian Reality*, 47.
21. Fierke, 153.
22. Rosenberg, "What Matters: The Role of Values in Transformations toward Sustainability: A Case Study of Coffee Production in Burundi."
23. Sharma.
24. United Nations, *United Nations Universal Declaration of Human Rights*.
25. Sharma.
26. Stapp, 153.
27. Wendt, 174.
28. Zanotti, *Ontological Entanglements, Agency and Ethics in International Relations*, 66.
29. Ibid., 75.
30. Barad.
31. Ibid., 338.
32. Ibid., 336.
33. Ibid., 393.
34. Ibid.

CHAPTER 9: FRACTALS

1. Bohm and Peat.
2. See Perey, "Organizing Sustainability and the Problem of Scale: Local, Global, or Fractal" and Sharma, *Radical Transformational Leadership*.
3. Bohm, *Wholeness and the Implicate Order*, 20-21.
4. Lenton, "Environmental Tipping Points."
5. Bohm, 24.
6. Bohm and Peat, 151.
7. Bohm.
8. Adnan, *The Politics of Waking Up*, 44.
9. Downton, *Ecopolis: Architecture and Cities for a Changing Climate*, 24.
10. Ibid., 27.
11. Hamilton, M., *Integral City: Evolutionary Intelligences for the Human Hive*, 67.
12. Sharma.

13. Rovelli, *The Order of Time*.
14. Wendt, 128 and 180 (italics added).
15. Wendt's ideas are based on time-symmetric interpretations of quantum theory, which hold that "fundamental physical principles are symmetric under time-reversal, meaning that in closed systems their equations can be solved forwards or backwards" (p. 179).
16. Wraight, "Non-Native Discourse about the Goals of the Onondaga Nation's Land Rights Action."
17. Rovelli, 134 (emphasis in original).
18. Whitehead, "Process and Reality: An Essay in Cosmology."
19. Roberts, "From Things to Events: Whitehead and the Materiality of Process," 972.
20. Clayton, "Introduction to Process Thought," 9.
21. Whitehead, 585.
22. Eastman and Keeton, *Physics and Whitehead: Quantum, Process and Experience*
23. Stapp.
24. Whitehead, "Modes of Thought," 871.
25. Jungerman, "Evidence for Process in the Physical World."
26. Ibid., 53.
27. Zhang, "The World's Fastest Camera Can Shoot 10 Trillion Frames Per Second."
28. Stapp, 33.
29. Fuller, *I Seem To Be a Verb*.
30. See Campbell, *The Hero With a Thousand Faces*.
31. Moore et al. "Scaling Out, Scaling Up, Scaling Deep: Strategies of Non-profits in Advancing Systemic Social Innovation."
32. See McCaffrey and Bhowmik, "Countdown to Drawdown: an initial overview of exponential scaling of potential societal tipping points for deep decarbonization of global energy infrastructure by 2050" and Bhowmik et al., "Powers of 10: Seeking 'Sweet Spots' for Rapid Climate and Sustainability Actions between Individual and Global Scales."
33. Ibid.
34. Adnan.
35. Wendt, 265.
36. Ibid.
37. Fierke and Antonio-Alfonso, "Language, Entanglement and the New Silk Roads," 19.
38. Bohm, *On Dialogue*, 7.
39. Ibid., 2.
40. Downton, 28.
41. Jadczyk, *Quantum Fractals: From Heisenberg's Uncertainty to Barnsley's Fractality*, 6.
42. Sharma.
43. Bohm and Peat, 56.
44. Barad, 396.

CHAPTER 10: YOU

1. Sharma, *Radical Transformational Leadership.*
2. O'Brien and Sygna, "Responding to Climate Change: The Three Spheres of Transformation" and O'Brien, "Is the 1.5°C Target Possible? Exploring the Three Spheres of Transformation."
3. Bhowmik et al.
4. McCaffrey, "From Person to Planet in Ten Easy Steps."
5. Bohm, *Wholeness and the Implicate Order*, 24.
6. Shove, "Putting Practice into Policy: Reconfiguring Questions of Consumption and Climate Change," 418.
7. Shove et al. *The Dynamics of Social Practice: Everyday Life and how it Changes*, 14.
8. Wraight, 2008.
9. The Alternative UK, "The Values we Use."
10. Sharma, 4.
11. Fierke, 152.
12. Meadows, 19.
13. Roberts, 981.
14. Zohar, *The Quantum Leader: A Revolution in Business Thinking and Practice*, 26.
15. Barad, 380 (italics added).
16. Thorne and Zurek, "John Archibald Wheeler: A Few Highlights of His Contributions to Physics," 35. The full quote is as follows: "if one really understood the central point and its necessity in the construction of the world, one ought to be able to state it [the relation between the quantum and measurement] in one clear, simple sentence. Until we see the quantum principle with this simplicity we can well believe that we do not know the first thing about the universe, about ourselves, and about our place in the universe."

EPILOGUE: QUESTIONS

1. Bohm, *Wholeness and the Implicate Order*, 36.
2. Wendt.
3. Nowotny, *The Cunning of Uncertainty*, 172.
4. For example, see Lieberman, *Social: Why Our Brains Are Wired to Connect*; Christakis and Fowler, *Connected: The Amazing Power of Social Networks and How they Shape our Lives.*
5. For example, see Brown and Harris, *The Human Capacity for Transformational Change: Harnessing the Collective Mind*; de la Cadena and Blaser, *A World of Many Worlds.*
6. Wendt.
7. El Khoury, *Globalization Development and Social Justice: A Propositional Political Approach*, 207.
8. The significance of quantum physics for geography was raised by William Peterman in 1994 in his article titled "Quantum Theory and Geography: What Can Dr. Bertlmann Teach Us?" Peterman is interested in how quantum theory can provide a scientific basis for human geography in a way that avoids the limitations of positivist science and recognizes its particular

relevance to understanding human-environment relationships in ways that are not deterministic.

9.   Freire, *Pedagogy of the Oppressed*, 139.
10. Escobar, *Pluriversal Politics: The Real and the Possible*.
11. Schneider-Mayerson and Bellamy, 8.
12. Rifkin, *The Empathic Civilization: The Race to Global Consciousness in a World in Crisis*, 616.
13. Wright, *Envisioning Real Utopias*.
14. El Khoury.
15. Baert, 156.
16. Worster, *Nature's Economy: The History of Ecological Ideas*, 388.
17. James, "Pragmatism."
18. von Baeyer, 205 (italics in original).
19. Lynch and Veland, *Urgency in the Anthropocene*, 4.
20. Sharma.
21. Foster, *The Sustainability Mirage: Illusion and Reality in the Coming War on Climate*, 56.
22. Hawken, *Blessed Unrest How the Largest Social Movement in History Is Restoring Grace, Justice, and Beauty to the World*.
23. El Khoury.
24. Barad.
25. Klein, *This Changes Everything: Capitalism vs. the Climate*.
26. Ghose, "How Quantum Physics Can Help Us Fight Climate Change."
27. Al-Khalili and McFadden, 295.

REFERENCES

Adnan, Indra. (2020) *The Politics of Waking Up*. London: Perspectiva Press.

Al-Khalili, Jim and McFadden, Johnjoe. (2014) *Life on the Edge: The Coming of Age of Quantum Biology*. London: Bantam Press.

Alaimo, Stacy. (2010) *Bodily Natures: Science, Environment, and the Material Self*. Bloomington: Indiana University Press.

Alexander, Stephon. (2016) *The Jazz of Physics: The Secret Link between Music and the Structure of the Universe*. New York: Basic Books.

Allison, Elizabeth A. 2015. "The Spiritual Significance of Glaciers in an Age of Climate Change." *Wiley Interdisciplinary Reviews: Climate Change* 6 (5): 493–508.

Alonso-Sanz, Ramon. (2019) *Quantum Game Simulation*. Cham, Switzerland: Springer.

Asch, Michael and Macklem, Patrick. (1991) "Aboriginal Rights and Canadian Sovereignty: An Essay on R. v. Sparrow." *Alberta Law Review* 29(2): 498-511.

Baert, Patrick. (2005) *Philosophy of the Social Sciences: Towards Pragmatism*. Cambridge: Polity.

Ball, Philip. (2018) *Beyond Weird: Why Everything You thought You Knew about Quantum Physics is Different*. London: Bodley Head.

Ball, Philip. (2006) *Critical Mass: How One Thing Leads to Another*. New York: Farrar, Straus and Giroux.

Baggerman, Ton. (2019) *It's About Us: Meaning, Emotions and Mental Health in Post-Newtonian Reality*. Dredggepress. Available at: https://tonbaggerman.nl/

Barad, Karen. (2007) *Meeting the Universe Halfway: Quantum Physics and the Entanglement of Matter and Meaning*. Durham: Duke University Press Books.

Bendell, Jem. (2018) "Deep Adaptation: A Map for Navigating Climate Tragedy." *IFLAS Occasional Paper* 2. Available at: https://www.lifeworth.com/deepadaptation.pdf.

Bennett, Jane. (2010) *Vibrant Matter: A Political Ecology of Things*. Durham: Duke University Press.

Berkes, Fikret. (2008) *Sacred Ecology*. Second Edition. New York: Routledge.

Berry, Thomas. (2014) "The New Story." Pages 12-16 in Mary Evelyn Tucker and John Grim (eds.), *Thomas Berry: Selected Writings on the Earth Community*. Maryknoll, NY: Orbis Books.

Bethell, Christina D., Solloway, Michele R., Guinosso, Stephanie, et al. (2017) "Prioritizing Possibilities for Child and Family Health: An Agenda to Address Adverse Childhood Experiences and Foster the Social and Emotional Roots of Well-being in Pediatrics." *Academic Pediatrics* Sep-Oct 2017:17(7S):S36-S50.

Bethell, Christina D, Gombojav, Narangerel, and Whitaker, Robert C. (2019)

"Family Resilience and Connection Promote Flourishing Among US Children, Even Amid Adversity." *Health Affairs* (Millwood) 38(5):729-737.

Bhaskar, Roy, Esbjörn-Hargens, Sean, Hedlund, Nicholas, and Hartwig, Mervyn Hartwig (eds.), (2016) *Metatheory for the Twenty-First Century: Critical Realism and Integral Theory in Dialogue.* London: Routledge.

Bhowmik, Avit K. et al. (2020) Powers of 10: Seeking 'Sweet Spots' for Rapid Climate and Sustainability Actions between Individual and Global Scales." *Environmental Research Letters* 15 094011.

Bohm, David. (1996) *On Dialogue.* New York: Routledge.

Bohm, David. (1980) *Wholeness and the Implicate Order.* London: Routledge and Kegan Paul.

Bohm, David and Peat, F. David. (1987) *Science, Order and Creativity.* Toronto: Bantam.

Boyd, Emily, Boykoff, Maxwell and Newell, Peter. (2011). "The 'New' Carbon Economy: What's New?" *Antipode* 43 (3): 601–611.

Bradford, Louise. (2019) *Save the World: There is No Planet B. Things You Can Do Right Now to Save our Planet*, London, United Kingdom: Summersdale Publishers Ltd.

Bradley, Keith. (2011) *The Cambridge World History of Slavery, AD 1420 - AD 1804.* Vol. 3. Cambridge, UK: Cambridge University Press.

Brown, Lester R. 2009. *Plan B 4.0: Mobilizing to Save Civilization.* New York: W.W. Norton.

Brown, Steven R. (1993) "A Primer on Q Methodology." *Operant Subjectivity* 16(3/4): 91–138.

Brown, Valerie A. and Harris, John A. (2014) *The Human Capacity for Transformational Change: Harnessing the Collective Mind.* Abingdon: Earthscan/Routledge.

Burgess, J. Peter. (2018). "Science Blurring Its Edges into Spirit: The Quantum Path to Ātma." *Millennium* 47 (1): 128–141.

Burwell, Jennifer. (2018) *Quantum Language and the Migration of Scientific Concepts.* Cambridge, Massachusetts: MIT Press.

Busemeyer, Jerome R. and Bruza, Peter D. (2012) *Quantum Models of Cognition and Decision.* Cambridge: Cambridge University Press.

Cameron, Emilie. (2012) "New Geographies of Story and Storytelling." *Progress in Human Geography* 36(5):573-592.

Campbell, Joseph. (2000) *The Hero With a Thousand Faces.* (Third Edition). Novato, California: New World Library.

Capra, Fritjof. (1991) *The Tao of Physics.* Third Edition. Boston: Shambhala.

Carroll, Sean. (2019) *Something Deeply Hidden: Quantum Worlds and the Emergence of Spacetime.* New York: Dutton.

Chen, Sophia. (2019) "Even Huge Molecules Follow the Quantum World's Bizarre Rules." *Wired.* September 23, 2019. https://www.wircd.com/story/even-huge-molecules-follow-the-quantum-worlds-bizarre-rules/

Christakis, Nicholas and Fowler, James. (2009) *Connected: The Amazing Power of Social Networks and How they Shape our Lives*. London: Harper Press.

Clayton, Philip. (2003) "Introduction to Process Thought," Pages 3-13 in Timothy E. Eastman and Hank Keeton (eds). *Physics and Whitehead: Quantum, Process and Experience*, Albany: State University of New York Press.

Coole, Diana, and Frost, Samantha (eds.). (2010) *New Materialisms: Ontology, Agency, and Politics*. Durham NC: Duke University Press.

Crease, Robert P. and Goldhaber, Alfred Scharff. (2014) *The Quantum Moment: How Planck, Bohr, Einstein, and Heisenberg Taught Us to Love Uncertainty*. New York: W.W. Norton & Company.

Danilov, Vladimir V. and Lambert-Mogiliansky, Ariane. (2018) "Preparing a (Quantum) Belief System." *Theoretical Computer Science*, 752: 97–103.

de la Cadena, Marisol and Blaser, Mario (eds.). (2018) *A World of Many Worlds*. Durham, NC: Duke University Press.

Deloria, Vine Jr. (2012) *The Metaphysics of Modern Existence*. Golden, Colorado: Fulcrum.

Dennett, Daniel C. (2003) *Freedom Evolves*. New York: Viking.

Der Derian, James and Wendt, Alexander. (2020) " 'Quantizing International Relations': The Case for Quantum Approaches to International Theory and Security Practice." *Security Dialogue* 51(5): 399–413.

Downton, Paul F. (2009) *Ecopolis: Architecture and Cities for a Changing Climate*. Dordrecht and Collingwood: Springer and CSIRO Publishing.

Doyle, Samuel H., Hubbard, Alun, van de Wal, Roderik S. W., et al. (2015) "Amplified Melt and Flow of the Greenland Ice Sheet Driven by Late-summer Cyclonic Rainfall." *Nature Geoscience* 8(8): 647-653.

Eastman, Timothy E. and Keeton, Hank (eds.). (2003) *Physics and Whitehead: Quantum, Process and Experience*, Albany: State University of New York Press.

Eisenstein, Charles. (2018) *Climate: A New Story*. Berkeley: North Atlantic Books.

El Khoury, Ann. (2014) *Globalization Development and Social Justice: A Propositional Political Approach*. London: Routledge.

Esbjörn-Hargens, Sean. (2010) "An Ontology of Climate Change: Integral Pluralism and the Enactment of Multiple Objects." *Journal of Integral Theory and Practice* 5(1):143-174.

Escobar, Arturo. (2020) *Pluriversal Politics: The Real and the Possible*. Durham, NC: Duke University Press.

Escobar, Arturo. (2018) *Designs for the Pluriverse: Radical Inetrdependence, Autonomy, and the Making of Worlds*. Durham, NC: Duke University Press.

Esposito, John. (2019). "In Lak'ech—A la K'in" Pages 132-140 in Matthew Schneider-Meyerson and Brent Ryan Bellamy (eds) *An Ecotopian Lexicon*. Minneapolis: University of Minnesota Press.

Fein, Yaakov Y., Geyer, Philipp, Zwick, Patrick, and Kialka, Filip. (2019) "Quantum Superposition of Molecules Beyond 25 kDa." *Nature Physics* 15(12): 1-4.

Feola, Giuseppe. (2015). "Societal Transformation in Response to Global

Environmental Change: A Review of Emerging Concepts." *Ambio* 44 (5): 376–90.

Ferrari, Pier F. and Coudé, Gino. (2018) "Mirror Neurons, Embodied Emotions, and Empathy." Pages 67-77 in K.Z. Meyza and E. Knapska (eds.), *Neuronal Correlates of Empathy: From Rodents to Humans*. London: Academic Press.

Fierke, Karin M. (2017) "Consciousness at the Interface: Wendt, Eastern Wisdom and the Ethics of Intra-Action." *Critical Review* 29 (2): 141–69.

Fierke, Karin M. and Antonio-Alfonso, Fransciso. (2018) "Language, Entanglement and the New Silk Roads." *Asian Journal of Comparative Politics* 3 (3): 14-206.

Figueres, Christiana and Tom Rivett-Carnac. (2020) *The Future We Choose: Surviving the Climate Crisis*. New York: Alfred A. Knopf.

Foster, John. (2008). *The Sustainability Mirage: Illusion and Reality in the Coming War on Climate*. London: Earthscan.

Foust, Christina R. and Murphy, William O'Shannon. (2009) "Revealing and Reframing Apocalyptic Tragedy in Global Warming Discourse." *Environmental Communication* 3(2):151-167.

Franzen, Jonathan. (2019) "What if we stopped pretending?" *The New Yorker Magazine*, September 8, 2019. Available at: https://www.newyorker.com/culture/cultural-comment/what-if-we-stopped-pretending

Freire, Paolo. (1992) *Pedagogy of Hope*. London: Bloomsbury.

Freire, Paolo. (1970) *Pedagogy of the Oppressed*. New York: Continuum.

Frischmann, Chad, Mehra, Mamta, Allard, Ryan, et al. (2020) "Drawdown's 'System of Solutions' Helps to Achieve the SDGs" in W. Leal Filho et al. (eds.), *Partnerships for the Goals*, Encyclopedia of the UN Sustainable Development Goals. Cham: Springer Nature.

Fuchs, Christopher A. (2017) "On Participatory Realism." Pages 113-134 in Ian T. Durham and Dean Rickles (eds.) *Information and Interaction: Eddington, Wheeler, and the Limits of Knowledge*. Cham: Springer.

Fuchs, Christopher A. and Stacey, Blake C. (2019) "QBism: Quantum Theory as a Hero's Handbook. *Proceedings of the International School of Physics "Enrico Fermi"*: Course 197, Foundations of Quantum Theory. Available at: https://arxiv.org/pdf/1612.07308.pdf

Fuller, R. Buckminster (with Jerome Agel and Quenten Fiore). (1970) *I Seem to Be a Verb*. New York. Bantam Books, Inc.

Ghose, Shohini. (2015) "How Quantum Physics Can Help us Fight Climate Change." *TEDx Victoria*, 2015. Available at: https://www.youtube.com/watch?v=x02ZKWkexeo

Gilder, Louisa. (2009) *The Age of Entanglement: When Quantum Physics Was Reborn*. New York: Vintage.

Gisin, Nicolas. (2014). *Quantum Chance: Nonlocality, Teleportation and Other Quantum Marvels*. Cham, Switzerland: Springer.

Gram-Hanssen, Irmelin, Schafenacker, Nicole, and Bentz, Julia. (2021) "Decolonizing Transformations through 'Right Relations'." *Sustainability*

Science. Available at: https://doi.org/10.1007/s11625-021-00960-9

Griffin, Paul. (2017) *CDP Carbon Majors Report 2017.* CDP, July 2017 in partnership with the Climate Accountability Institute.

Gunderson, Lance H., and Holling, C. S. (eds.). (2002) *Panarchy: Understanding Transformations in Human and Natural Systems.* Washington: Island Press.

Haff, Peter. (2014) "Humans and Technology in the Anthropocene: Six Rules." *The Anthropocene Review* 1(2): 126–36.

Hamilton, Clive. (2014) *Earthmasters: The Dawn of the Age of Climate Engineering.* New Haven: Yale University Press.

Hamilton, Marilyn. (2008) *Integral City: Evolutionary Intelligences for the Human Hive.* Gabriola, Canada: New Society Publishers.

Haven, Emmanuel and Khrennikov, Andrei. (2013) *Quantum Social Science.* New York: Cambridge University Press.

Hawken, Paul (2017) *Drawdown: The Most Comprehensive Plan Ever Proposed to Reverse Global Warming.* New York, NY: Penguin Books.

Hawken, Paul. (2007) *Blessed Unrest: How the Largest Social Movement in History Is Restoring Grace, Justice, and Beauty to the World.* New York: Penguin Books.

Hickman, Leo. (2010) "James Lovelock: Humans are too stupid to prevent climate change." *The Guardian*, 29 March 2010. Available at: https://www.theguardian.com/science/2010/mar/29/james-lovelock-climate-change

Hobbs, Benjamin F. (1997) "Bayesian Methods for Analysing Climate Change and Water Resource Uncertainties." *Journal of Environmental Management* 49: 53-72.

Hochachka, Gail. (2019) "On Matryoshkas and Meaning-Making: Understanding the Plasticity of Climate Change." *Global Environmental Change* 57: 101917.

Hord, Fred Lee (Mzee Lasana Okpara), and Lee, Jonathan Scott (eds.). (2016) *I Am Because We Are: Readings in Africana Philosophy.* Revised Edition. Amherst & Boston: University of Massachusetts Press.

Horgan, John. (2018) *Mind-Body Problems: Science, Subjectivity & Who We Really Are.* Self-Published. Available at: https://mindbodyproblems.com/

Hübl, Thomas. (2020) *Healing Collective Trauma: A Process for Integrating Our Intergenerational and Cultural Wounds.* Boulder, CO: Sounds True.

IPBES. (2019) *Global Assessment Report on Biodiversity and Ecosystem Services of the Intergovernmental Science-Policy Platform on Biodiversity and Ecosystem Services.* Eduardo S. Brondizio, Josef Settele, Sandra Díaz, and Hien T. Ngo (editors). Bonn, Germany: IPBES Secretariat.

IPCC. (2018) *Global Warming of 1.5°C.* IPCC Special Report on the impacts of global warming of 1.5°C above pre-industrial levels and related global greenhouse gas emission pathways, in the context of strengthening the global response to the threat of climate change, sustainable development, and efforts to eradicate poverty. [V. Masson-Delmotte, et al (eds.)]. https://www.ipcc.ch/sr15/download/

Jackson, R.B., Le Quéré, C., Andrew, R.M., et al. (2019) *Global Energy Growth Is*

Outpacing Decarbonization. A special report for the United Nations Climate Action Summit, September 2019. Global Carbon Project, International Project Office, Canberra Australia.

Jackson, William Anthony (1999) Dualism, Duality and the Complexity of Economic Institutions. *International Journal of Social Economics* 26(4): 545-558.

Jadczyk, Arkadiusz. (2014) *Quantum Fractals: From Heisenberg's Uncertainty to Barnsley's Fractality.* New Jersey: World Scientific.

James, William. (1907) *Pragmatism.* Cambridge: Harvard University (A Public Domain Book).

Johnson, Mark. (1987) *The Body in the Mind: The Bodily Basis of Meaning, Imagination, and Reason.* Chicago: University of Chicago Press.

Joronen, Mikko and Häkli, Jouni. (2016) "Politicizing Ontology" *Progress in Human Geography* 41 (5): 561-579.

Jungerman, John A. (2003) "Evidence for Process in the Physical World," Pages 47-56 in Timothy E. Eastman and Hank Keeton (eds.). *Physics and Whitehead: Quantum, Process and Experience*, Albany: State University of New York Press.

Kahane, Adam (2017) *Collaborating with the Enemy: How to Work with People You Don't Agree with or Like or Trust.* Oakland, CA: Berrett-Koehler Publishers, Inc.

Kaiser, David. (2012) *How the Hippies Saved Physics: Science, Counterculture, and the Quantum Revival.* New York: W.W. Norton & Company.

Kegan, Robert, and Lahey, Lisa Laskow. (2009) *Immunity to Change: How to Overcome It and Unlock the Potential in Yourself and Your Organization.* Boston, Harvard Business Press.

Keller, Klaus. (2013). "Bayesian Decision Theory and Climate Change." Pages 1-14 in J.F. Shogren (ed.), *Encyclopedia of Energy, Natural Resources and Environmental Economics.* London: Elsevier Science.

Klein, Naomi. (2014) *This Changes Everything: Capitalism vs. the Climate.* New York: Simon and Schuster.

Kleppner, Daniel and Jackiw, Roman. (2000). "One Hundred Years of Quantum Physics" *Science* 289(5481):893-898.

Koch, Christof. (2019) *The Feeling of Life Itself: Why Consciousness is Widespread but Can't be Computed.* Cambridge, Massachusetts: The MIT Press.

Konvalinka, Ivana, Xygalatas, Dimitris, Bulbulia, Joseph, et al. (2011) "Synchronized Arousal Between Performers and Related Sectators in a Fire-walking Ritual." *Proceedings of the National Academy of Sciences of the United States of America* 108(20): 8514-8519.

Kotler, Shlomi, Peterson, Gabriel A., Shojaee, Ezad, et al. (2021) "Direct Observation of Deterministic Macroscopic Entanglement." *Science* 372: 622–625.

Kuhn, Thomas S. (1962) *The Structure of Scientific Revolutions.* Chicago: University of Chicago Press.

Lakoff, George, and Johnson, Mark. (1980) *Metaphors We Live By*. Chicago: University of Chicago Press.

Lakoff, George, and Johnson, Mark. (1999) *Philosophy in the Flesh: The Embodied Mind and Its Challenge to Western Thought*. New York: Basic Books.

Leichenko, Robin and O'Brien, Karen. (2019) *Climate and Society: Transforming the Future*. Cambridge: Polity.

Leichenko, Robin and O'Brien, Karen. (2008) *Environmental Change and Globalization: Double Exposures*. New York: Oxford University Press.

Lenton, Timothy M. (2013) "Environmental Tipping Points." *Annual Review of Environment and Resources* 38: 1-29.

Lieberman, Matthew D. (2013) *Social: Why Our Brains Are Wired to Connect*. New York: Broadway Books.

Lieberman, Matthew D. and Eisenberger, Naomi I. (2009) "Pains and Pleasures of Social Life." *Science* 323(5851): 890-891.

Lorenzoni, Irene and Hulme, Mike. (2013) "Believing Is Seeing: Laypeople's Views of Future Socio-Economic and Climate Change in England and in Italy." *Public Understanding of Science* 18: 383–400.

Lynch, Amanda H. and Veland, Siri. (2018). *Urgency in the Anthropocene*. Cambridge, Massachusetts: The MIT Press.

Lynch, Amanda H., Griggs, David, Joachim, Lee, and Walker, Jackie. (2013) "The Role of the Yorta Yorta People in Clarifying the Common Interest in Sustainable Management of the Murray-Darling Basin, Australia," *Policy Sciences* 46(2):109-123.

Marshall, George. (2015) *Don't Even Think About It: Why Our Brains Are Wired to Ignore Climate Change*. London: Bloomsbury.

Mathews, Freya. (2003) *For Love of Matter: A Contemporary Panpsychism*. Albany: SUNY Press.

McCaffrey, Mark. (2017) "From Person to Planet in Ten Easy Steps" *Medium* October 3, 2017. Available at: https://medium.com/@markmccaffrey_90684/from-person-to-planet-in-ten-easy-steps-29b08f13edb5

McCaffrey, Mark and Bhowmik, Avit. (2017) "Countdown to Drawdown: an initial overview of exponential scaling of potential societal tipping points for deep decarbonization of global energy infrastructure by 2050." EGU General Assembly, *Geophysical Research Abstracts* 19: EGU2017-17224.

Meadows, Donella H. (1999) "Leverage Points: Places to Intervene in a System." The Sustainability Institute. Available at: http://donellameadows.org/archives/leverage-points-places-to-intervene-in-a-system/

Meadows, Donella H., Randers, Jorgen, Meadows, Dennis L. and Behrens, William. (1972) *The Limits to Growth: A Report for the Club of Rome's Project on the Predicament of Mankind*. New York: Universe Books.

Mermin, N. David. (2014) "Physics: QBism Puts the Scientist Back Into Science." *Nature* 507 (7493): 421-423.

Milkoreit, Manjana. (2017) *Mindmade Politics: The Cognitive Roots of International*

Climate Governance. Cambridge, Massachusetts: The MIT Press.

Milman, Oliver. (2017) "Planet has just 5% chance of reaching Paris climate goal, study says" *The Guardian*, July 31, 2017. Available at: https://www.theguardian.com/environment/2017/jul/31/paris-climate-deal-2c-warming-study.

Mooney, Chris. (2015) "This is Climate Skeptics' Latest Argument about Melting Polar Ice – And Why it's Wrong." *The Washington Post*, May 27, 2015. Available at: https://www.washingtonpost.com/news/energy-environment/wp/2015/05/27/climate-skeptics-think-you-shouldnt-worry-about-melting-polar-ice-heres-why-theyre-wrong/

Moore, Michele-Lee, Riddell, Darcy J., and Vocisano, Dana. (2015). "Scaling Out, Scaling Up, Scaling Deep: Strategies of Non-profits in Advancing Systemic Social Innovation." *Journal of Corporate Citizenship* 58: 67-84.

Musser, George. (2016) "How Einstein Revealed the Universe's Strange "Nonlocality." *Scientific American*, November 1, 2015. Available at: https://www.scientificamerican.com/article/how-einstein-revealed-the-universe-s-strange-nonlocality/

Nagel, Thomas. (2012) *Mind and Cosmos: Why the Materialist Neo-Darwinian Conception of Nature Is Almost Certainly False.* New York: Oxford University Press.

Nagel, Thomas. (1989) *The View from Nowhere.* New York: Oxford University Press.

Nilsson, Nils J. (2014) *Understanding Beliefs.* Cambridge, Massachusetts: The MIT Press.

Nowotny, Helga. (2016) *The Cunning of Uncertainty.* Cambridge: Polity Press.

O'Brien, Karen. (2021) "Reflecting on the Anthropocene: The Call for Deeper Transformations." *Ambio*: https://doi.org/10.1007/s13280-020-01468-9

O'Brien, Karen. (2018) "Is the 1.5°C Target Possible? Exploring the Three Spheres of Transformation." *Current Opinion in Environmental Sustainability*, 31 (April): 153–160.

O'Brien, Karen. (2016) "Climate Change and Social Transformations: Is It Time for a Quantum Leap?" *Wiley Interdisciplinary Reviews: Climate Change*, 7: 618–626.

O'Brien, Karen, El Khoury, Ann, Schafenacker, Nicole, and Rosenfeld, Jordan (eds.) (2019) *Our Entangled Future: Stories to Empower Quantum Social Change.* AdaptationCONNECTS Research Project, University of Oslo, Norway. Available at: http://cchange.no/ourentangledfuture

O'Brien, Karen, and Sygna, Linda. (2013) "Responding to Climate Change: The Three Spheres of Transformation." Pages 16-23 n *Proceedings of Transformation in a Changing Climate.* Oslo, Norway: University of Oslo. Available at: https://bit.ly/3kQbRSl

Oliver, Tom. (2021) *The Self Delusion: The Surprising Science of Our Connection to Each Other and the Natural World.* London: Weidenfeld & Nicholson.

O'Neil, Cathy. (2016) *Weapons of Math Destruction: How Big Data Increases Inequality and Threatens Democracy.* New York: Crown Publishers.

Orrell, David. (2018) *Quantum Economics: The New Science of Money.* London: Icon Books.

Otto, Ilona M., Wiedermann, Mark, Cremades, Roger, et al. (2020). "Human Agency in the Anthropocene." *Ecological Economics* 167: 106463.

Perey, Robert. (2014) "Organizing Sustainability and the Problem of Scale: Local, Global, or Fractal." *Organization & Environment* 27(3) 215–222.

Peterman, William. (1994) "Quantum Theory and Geography: What Can Dr. Bertlmann Teach Us?" *The Professional Geographer* 46(1): 1-9.

Phillips, Jonathan. (2012) "Storytelling in Earth Sciences: The Eight Basic Plots." *Earth-Science Reviews* 115: 153–162.

Proietti, Massimiliano, Pickston, Alexander, Graffitti, Francesco, et al. (2019). "Experimental Test of Local Observer Independence." *Science Advances* 5 (9): eaaw9832.

Raftery, Adrian E., Zimmer, Alec, Frierson, Dargan M.W., et al. (2017) "Less than 2°C Warming by 2100 Unlikely." *Nature Climate Change* 7: 637-641.

Rainforest Action Network, et al. (2021) *Banking on Climate Chaos: Fossil Fuel Finance Report 2021.* Available at: https://www.ran.org/wp-content/uploads/2021/03/Banking-on-Climate-Chaos-2021.pdf

Riedy, Chris. (2019) *The Witnesses.* Pages 15-29 in K. O'Brien et al. (eds.) *Our Entangled Future: Stories to Empower Quantum Social Change.* AdaptationCONNECTS Research Project, University of Oslo, Norway. Available at: http://cchange.no/ourentangledfuture

Rifkin, Jeremy. (2009) *The Empathic Civilization: The Race to Global Consciousness in a World in Crisis.* New York: Jeremy P. Tarcher/Penguin.

Roberts, J. Timmons, and Parks, Bradley. (2006) *A Climate of Injustice: Global Inequality, North-South Politics and Climate Policy.* Cambridge, Massachusetts: The MIT Press.

Roberts, Tom. (2014) "From Things to Events: Whitehead and the Materiality of Process." *Environment and Planning D: Society and Space* 32(6): 968-983.

Rockström, Johan, Gaffney, Owen, Rogelj, Joeri, et al. 2017. "A Roadmap for Rapid Decarbonization." *Science* 355(6331): 1269–71.

Rosenberg, Milda Nordbø. (2021). "What Matters: The Role of Values in Transformations toward Sustainability: A Case Study of Coffee Production in Burundi." *Sustainability Science.* https://doi.org/10.1007/s11625-021-00974-3

Rosiek, Jerry Lee, Snyder, Jimmy, and Pratt, Scott L. (2020) "The New Materialisms and Indigenous Theories of Non-Human Agency: Making the Case for Respectful Anti-Colonial Engagement," *Qualitative Inquiry* 26(3-4): 331-346.

Rosol, Christoph, Nelson, Sara, and Renn, Jürgen. (2017) "Introduction: In the Machine Room of the Anthropocene." *The Anthropocene Review* 4 (1): 2–8.

Rovelli, Carlo. (2018) *The Order of Time*. Translated by Erica Segre and Simon Carnell. United Kingdom: Allen Lane.

Rovelli, Carlo. (2016) *Reality Is Not What It Seems: The Journey to Quantum Gravity*. Translated by Simon Carnell and Erica Segre. New York: Riverhead Books.

Rovelli, Carlo. (2015) *Seven Brief Lessons on Physics*. Translated by Simon Carnell and Erica Segre. Great Britain: Penguin Books.

Samson, Paul R. and Pitt, David (eds.). (1999) *The Biosphere and Noosphere Reader: Global Environment, Society and Change*. London: Routledge.

Sapir, Edward. (1968). *The Selected Writings of Edward Sapir*. Fifth Printing. Berkeley: University of California Press.

Schlitz, Marilyn, Vieten, Cassandra, and Miller, Elizabeth M. (2010) "Worldview Transformation and the Development of Social Consciousness." *Journal of Consciousness Studies* 17(7-1):18-36.

Schneider-Mayerson, Matthew and Bellamy, Brent Ryan (eds.) (2019) *An Ecotopian Lexicon*. Minneapolis: University of Minnesota Press.

Scoville-Simonds, Morgan. (2018) "Climate, the Earth, and God – Entangled Narratives of Cultural and Climatic Change in the Peruvian Andes." *World Development* 110: 345-359.

Sharma, Monica. (2017) *Radical Transformational Leadership: Strategic Action for Change Agents*. Berkeley: North Atlantic Books.

Sharma, Monica. (2007) "From Personal to Planetary Transformation." *Kosmos* Fall/Winter 2007, 31-25.

Shaw, Christopher and Nerlich, Brigitte. (2015) "Metaphor as a Mechanism of Global Climate Change Governance: A Study of International Policies, 1992–2012." *Ecological Economics* 109:34-40.

Shove, Elizabeth. (2014) "Putting Practice into Policy: Reconfiguring Questions of Consumption and Climate Change." *Contemporary Social Science: Journal of the Academy of Social Sciences* 9(4): 415-429.

Shove, Elizabeth, Pantzar, Mika, and Watson, Matt. (2012) *The Dynamics of Social Practice: Everyday Life and how it Changes*. London: Sage.

Siegel, Daniel J. (2021) *IntraConnected: MWe (Me + We) as the Integration of Belonging and Identity*. New York: W.W. Norton & Company.

Siegel, Daniel J. (2007) *The Mindful Brain: Reflection and Attunement in the Cultivation of Well-Being*. New York: W.W. Norton & Company.

Stapp, Henry P. (2011) *Mindful Universe: Quantum Mechanics and the Participating Observer*. Heidelberg: Springer Science & Business Media.

Steffen, Will, Rockström, Johan, Richardson, Katherine, et al. (2018) "Trajectories of the Earth System in the Anthropocene." *Proceedings of the National Academy of Sciences* 115 (33): 8252–59.

Steffen, Will, Richardson, Katherine, Rockstrom, Johan, et al. (2015) "Planetary Boundaries: Guiding Human Development on a Changing Planet. *Science*, 347(6223), 1259855–1259855.

Stoknes, Per Espen. (2015) *What We Think About When We Try Not To Think About*

Global Warming. White River Junction: Chelsea Green Publishing.

Straw, Ed and Ison, Roy. (2018) "Duality, dualism, duelling and Brexit." *Open Democracy*, 1 March 2018. Available at: https://www.opendemocracy.net/en/can-europe-make-it/duality-dualism-duelling-and-brexit/

Taylor, Cynthia and Dewsbury, Bryan M. (2018) "On the Problem and Promise of Metaphor Use in Science and Science Communication," *Journal of Microbiology & Biology Education* 19(1):1-5.

Tebaldi, Claudia, Smith, Richard L., Nychka, Doug, and Mearns, Linda O. (2005) "Quantifying Uncertainty in Projections of Regional Climate Change: A Bayesian Approach to the Analysis of Multimodel Ensembles." *Journal of Climate* 18(10): 1524-1540.

Teilhard de Chardin, Pierre. (1959) *The Phenomenon of Man.* New York: Harper Perennial Modern Thought.

The Alternative UK (2021) "The Values We Use." Webpage. Available at: https://www.thealternative.org.uk/the-values-we-use

The BIG Bell Test Collaboration. (2018) "Challenging Local Realism with Human Choices." *Nature* 557:212–216.

The Shift Project. (2019) *Lean ICT: Towards Digital Sobriety.* Available at: https://theshiftproject.org/wp-content/uploads/2019/03/Lean-ICT-Report_The-Shift-Project_2019.pdf

Thorne, Kip S. and Zurek, Wojciech H. (2010) "John Archibald Wheeler: A Few Highlights of His Contributions to Physics," Pages 29-38 in Ignazio Ciufolini and Richard A. Matzner (eds.), *General Relativity and John Archibald Wheeler.* Dordrecht: Springer.

Tononi, Giulio and Koch, Christof. (2015) "Consciousness: Here, There and Everywhere?" *Philosophical Transactions of the Royal Society B: Biological Sciences* 370 (1668): 20140167–20140167.

Torey, Zoltan. (2014) *The Conscious Mind.* Cambridge, Massachusetts: The MIT Press.

United Nations. (1948) *United Nations Universal Declaration of Human Rights.* Available at: https://www.un.org/en/universal-declaration-human-rights/

van der Kolk, Bessel. (2015) *The Body Keeps the Score: Brain, Mind, and Body in the Healing of Trauma.* New York: Penguin Books.

von Baeyer, Hans Christian. (2016) *QBism: The Future of Quantum Physics.* Cambridge, Massachusetts: Harvard University Press.

Waddock, Sandra, and Kuenkel, Petra. (2019) "What Gives Life to Large System Change?" *Organization & Environment*, May, 108602661984248.

Watts, Simon and Stenner, Paul. (2012) *Doing Q Methodological Research: Theory, Method and Interpretation.* London: Sage.

Watts, Simon and Stenner, Paul. (2003) "Q Methodology, Quantum Theory, and Psychology." *Operant Subjectivity* 26(4):157-175.

Weber, Andreas. (2013) *Enlivenment: Towards a Fundamental Shift in the Concepts of Nature, Culture and Politics.* Berlin: Heinrich Böll Stiftung.

Wendt, Alexander. (2015) *Quantum Mind and Social Science: Unifying Physical and Social Ontology.* Cambridge, U.K.: Cambridge University Press.

Whitehead, Alfred North. (1953) "Process and Reality: An Essay in Cosmology." Pages 562-746 in F.S.C. Northrup and Mason W. Gross: *Alfred North Whitehead: An Anthology.* Cambridge, U.K.: Cambridge University Press.

Whitehead, Alfred North. (1953) "Modes of Thought." Pages 857-924 in F.S.C. Northrup and Mason W. Gross: *Alfred North Whitehead: An Anthology.* Cambridge, U.K.: Cambridge University Press.

Wilber, Ken. (2000) *Sex, Ecology, Spirituality: The Spirit of Evolution.* The Collected Works of Ken Wilber, Volume Six (Second, Revised Edition). Boston: Shambhala.

Wilber, Ken (ed.). (2011) *Quantum Questions: Mystical Writings of the World's Great Physicists.* Boulder, CO; New York: Shambhala; Random House.

Wilber, Ken. (2001) *Eye to Eye: The Quest for the New Paradigm.* Boston: Shambhala.

Wilkerson, Isabel. (2020) *Caste: The Origins of Our Discontents.* New York: Random House.

Worster, Donald. (1994) *Nature's Economy: The History of Ecological Ideas.* Second Edition. Cambridge, U.K.: Cambridge University Press.

Wraight, Sarah. (2008) "Non-Native Discourse about the Goals of the Onondaga Nation's Land Rights Action" *Syracuse University Honors Program Capstone Projects.* 536. Available at: https://surface.syr.edu/honors_capstone/536

Wright, Eric Olin. (2010) *Envisioning Real Utopias.* London: Verso.

Xie, Jierui, Sreenivasan, S., Korniss, G., et al. (2011) "Social Consensus through the Influence of Committed Minorities." *Physical Review E* 84 (1): 011130.

Yunkaporta, Tyson. (2020) *Sand Talk: How Indigenous Thinking Can Save the World.* New York: HarperOne.

Zanotti, Laura. (2020) "De-colonizing the Political Ontology of Kantian Ethics: A Quantum Perspective," *Journal of International Political Theory.* Available at: https://doi.org/10.1177/1755088220946777

Zanotti, Laura. (2017) *Ontological Entanglements, Agency and Ethics in International Relations: Exploring the Crossroads.* London: Routledge.

Zhang, Michael. (2018) "The World's Fastest Camera Can Shoot 10 Trillion Frames Per Second." *Peta Pixel,* October 15, 2018. Available at: https://petapixel.com/2018/10/15/the-worlds-fastest-camera-can-shoot-10-trillion-frames-per-second/

Zohar, Danah. (2016) *The Quantum Leader: A Revolution in Business Thinking and Practice.* Amherst, NY: Prometheus Books.

Zohar, Danah, and Marshall, Ian N. (1994) *The Quantum Society: Mind, Physics, and a New Social Vision.* New York: Quill, 1994.

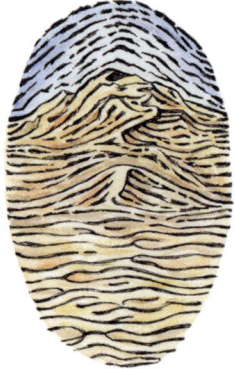

INDEX

ABOUT

KAREN O'BRIEN is a Professor in the Department of Sociology and Human Geography at the University of Oslo, Norway. She is also co-founder of cCHANGE, an organization that supports transformation in a changing climate. Karen's research emphasizes the social and human dimensions of climate change and implications for human security. Her current research focuses on the relationship between climate change adaptation and transformations to sustainability, with an emphasis on the role of creativity, collaboration, empowerment, and narratives. She is particularly interested in the role of beliefs, values, worldviews, and paradigms in generating conscious social change. Karen has participated in four reports for the Intergovernmental Panel on Climate Change, and as part of the IPCC, she was a co-recipient of the 2007 Nobel Peace Prize. She was named by Web of Science as one of the world's most influential researchers of the past decade in 2019 and 2020. In 2021 she was co-recipient of the BBVA Foundation Frontiers of Knowledge Award for Climate Change. Together with Robin Leichenko, she is the co-author of *Climate and Society: Transforming the Future*. Karen is passionate about engaging people with transformations to sustainability to realize our shared potential to thrive.

TONE BJORDAM is a Norwegian artist who specializes in projects related to nature, perceptions, and science. She works with video, animation films, nature photography, abstract and nature-inspired paintings, intricate, detailed drawings, and sculpture installations. Tone is passionate about finding ways to communicate science through art. www.tonebjordam.com

CHRISTINA BETHELL is Professor of Public Health and Human Development in the Bloomberg School of Public Health at Johns Hopkins University. For the past 35 years she has built her work and career around an intentional goal to catalyze health care and public health transformation at the policy, systems and practice levels. Christina is passionate about promoting self- and community-led healing and health and addressing childhood trauma.